The Hypnobehavioral Treatment of Eating and Anxiety Disorders

Donald Keppner, PhD

Scientific based theory and treatment

Anorexia Nervosa

Bulimia Nervosa

Post Traumatic Stress Disorder

Anxiety Disorders

Copyright © 2015 Donald R. Keppner, PhD

All rights reserved. This book or any portion thereof may not be reproduced or used in any manner whatsoever without the express written permission of the author except for the use of brief quotations.

First Printing, 2015

ISBN-13: 978-1511500227

ISBN-10: 1511500220

DEDICATION

This volume is dedicated to Vivian Hanson Meehan for her humanitarian concern, pioneering spirit and selfless dedication in founding and leading the first organization in America dedicated to alleviating the isolation, loneliness and suffering of people with eating disorders and their families; for her leadership in developing extensive programs which have impacted the lives of millions; and for her loving support of thousands of people on an individual basis.

ACKNOWLEDGMENTS

I wish to express my profound appreciation to everyone who helped to bring this book into existence. I also appreciate the ongoing effort to enable suffering people, their families and loved ones to cope with and overcome eating and anxiety disorders.

My wife Victoria has been a constant helper, supporter and friend.

Vivian Hanson Meehan first interested me in understanding and treating eating disorders. With her inquiring mind, she first grasped the immense value of including trance therapy in my treatment protocol.

Many others have my gratitude for their important contributions to the field and to my own progress.
These include Carla Arnell, Mike Alonzi,
Eugene Beeler, Jeff Baliki, Catherine Cotter MSW,
Valerie Gutmann, Dee Hubbard, Mary Parrent,
Jean Rubel ThD and Martin Ross.

The support and encouragement of Working Together for Humanity, Inc. is greatly appreciated.

DONALD R. KEPPNER, PH.D.

THE HYPNOBEHAVIORAL TREATMENT OF EATING AND ANXIETY DISORDERS
by Donald R. Keppner, Ph.D.

TABLE OF CONTENTS

I. INTRODUCTION — 7

II. BEHAVIORAL AND HYPNOTIC CONCEPTS COMMON TO EFFECTIVE PSYCHOTHERAPY — 8

    A. Respondent, Vicarious, and Operant Conditioning — 16

    B. Two-Process Theory of Avoidance Learning — 23

    C. Extinction, Spontaneous Recovery, and the Unconscious Mind — 26

    D. Abreactive Extinction and Hypnotherapy — 28

    E. Behavioral Model for Obsessive-Compulsive Disorders — 44

    F. In-Patient Treatment of Eating Disorders and Family Therapy — 53

    G. Eating Disorders: The Hypnobehavioral Model — 62

    H. Cognitive-Behavioral Treatment of Depression — 66

    I. State Dependent Learning and Generalization — 79

    J. Anxiety, Dissociation, and Conversion Features of Anorexia Nervosa and Bulimia — 82

    K. Trance and Separation Therapy — 119

    L. Hypnobehavioral Model of Obsessive-Compulsive Behavior and Unconscious Motivation — 129

    M. Bulimic Cognitive-Behavioral Feedback Loop — 160

    N. Spiritual Aspects of Psychotherapy — 167

    O. Laws for Successful Living — 175

P. Initial Interview, History Taking, and Selection of Clients	185
Q. Important Non-Psychological Variables	199
R. Post Traumatic Stress Disorder (PTSD)	222
Post Traumatic Stress Disorder PTSD Case Study 1	226
Post Traumatic Stress Disorder PTSD Case Study 2	230
Post Traumatic Stress Disorder PTSD Case Study 3	233
III. REFERENCES	238

The greatest power that humans possess is the ability to change their attitudes and values and formulate a philosophy of living and practical habits that leads to a rewarding lifestyle through hard work.

INTRODUCTION

The purpose of this book is to serve as a complete guide to the treatment of chronic eating disorders and chronic anxiety disorders. I emphasize *chronic* because most of my cases had failed in prior treatment and all had their problems for over three years. Reviews concerning the efficacy of family therapy and cognitive behavioral therapy for the treatment of this particular population show that these therapies are not very effective. All the cases I treated between 1975 and 1993 were from this population.

Exact figures cannot be found as the follow-up records on my cases are no longer available. However, two extremely reliable sources who followed my cases reported a 49% total cure, many after only three full days of treatment and on a six month follow-up. Over 70% rated themselves significantly helped; again, many after only a three-day treatment period on a six month follow-up.

II: BEHAVIORAL AND HYPNOTIC CONCEPTS COMMON TO EFFECTIVE PSYCHOTHERAPY

This section outlines the etiology and treatment of obsessive-compulsive disorders (OCD), anorexia-bulimia (A-B), and anxiety disorders (AD). It has been my experience that A-B and AD, including phobias, OCDs, and addictive disorders share common features that can be effectively controlled or eliminated by a combination of well understood behavioral and hypnotic procedures. The first objective is to describe the basic behavioral and hypnotic concepts that form the basis of all effective psychotherapy. These concepts and methods establish the foundation for understanding the etiology and treatment of chronic A-Bs, OCDs and ADs. The second objective is to explain how A-Bs and ADs are treated effectively by the proper combination of these techniques and to illustrate the methods through case examples.

Most of the supporting data for the hypotheses in this book are from empirical observations, successful clinical cases, and control group studies. Effort has been made to report and review substantiating experimental evidence, but my emphasis is on concepts, ideas, and methods. Specific studies are discussed only to the extent that they are critical or illustrate important concepts. Many of the ideas concerning dissociation, hypnosis, and fragmentation of personality are a modern restatement of late 19th-century hypnotists such as Pierre Janet (1906), who was among the first to record these phenomena. These valuable insights were unfortunately ignored with the abandonment of hypnosis as a major therapeutic tool by Freud and his followers and by later non-directive therapists.

Freudians and insight-oriented therapies based on the medical model do not offer a reasonable etiological explanation of A-B and AD, and have little therapeutic value (Eysenck, 1952; Hobbs, 1962; London, 1969; Dawes, 1994). I have found along with many others that behavioral therapy and cognitive behavior therapy (CBT) are more effective treatment approaches for A-Bs and ADs.

Much research has been done since 1982 when I wrote the first edition of this book which supports my treatment approach. However, my approach of treating the dissociative aspects of A-B and AD along with the cognitive-behavioral aspects has been seldom appreciated in the literature.

New treatment methods are often created based on old theories; however, many methods may have little to do with the theories. When one observes the application of these methods and ignores the theoretical suppositions, it becomes obvious that effective therapies work because they use cognitive-behavioral and hypnotic principles and procedures in an unsystematic manner.

Even though the major emphasis of this book is the treatment of anorexia and bulimia, this treatment is paradigmatic for the treatment of all anxiety disorders, obsessive-compulsive disorders, addictive-compulsive disorders, and dissociative disorders. The basic tenet of my approach is the behavioral approach, which regards A-B and AD behavior as maladaptive behaviors that are learned or conditioned and are, therefore, extinguishable and replaceable by more adaptive behaviors. When a complete learning history is obtained through the interview and hypnotic regression, the important learning experiences and reinforcers for the maladaptive behaviors become evident. Therefore with A-Bs and ADs, the basic processes determining the maladaptive overt and covert behaviors are understandable. The combination of classically conditioned emotional responses (CER), operant behavior, reinforcement or punishment, and vicarious conditioning explain the maladaptive emotional and behavioral patterns that characterize A-Bs and ADs.

A-B is more common in females than males. Eating disorders may begin in early childhood or adulthood, but usually begin in adolescence or pre-adolescence. Anorexics are usually more emaciated than bulimics; however, bulimics may possess body types that range from obese to emaciated. A-B is a disorder of thought and behavior that may or may not result in extreme weight fluctuations. Therefore, criterion concerning the percentage of weight either lost or gained may not be meaningful for understanding or classifying these disorders. Some bulimics maintain a normal weight by extreme binging and vomiting. What mediates A-Bs' behavior is a strong fear of weight gain along with similar cognitions and beliefs. Of course, years of starvation, binge-eating and vomiting, etc., wreak havoc on the entire body.

To summarize, A-Bs may lose their teeth and abuse laxatives and diuretics, causing severe electrolytic imbalances, irreversible physical damage, and death. Anorexics of course experience all the physiological consequences of starvation, including loss of brain mass and lowering of IQ. This loss of brain mass can cause a chronic brain syndrome that renders the anorexic untreatable. They simply cannot function well enough mentally to understand and benefit from treatment. Their short-term memory and attention is so impaired that they have extreme difficulty retaining what has been taught. These clients may be treatable after they have been medically (physiologically) stabilized as a reasonably well-functioning brain is necessary for learning.

A-Bs are generally compulsive, prone to mood fluctuations, and always obsessed with the fear of gaining weight. This fear, along with a variety of other negative emotions including guilt and anger, mediate their distorted body image (seeing themselves as fat even when emaciated or normal weight) or other conversion reactions. A-Bs experience the same conversion reactions, which include feelings of bloating in the abdomen, face and thighs, and a full feeling following the ingestion of even a small amount of food. These conversion reactions are often elicitable by simply touching a piece of food to their lips. Most become extremely agitated following the ingestion of almost any food or liquid. Gaining weight is perceived as worse than death, and even the thought of weight gain elicits extreme anxiety, depression, and often conversion reactions. Actual weight gain immediately mediates a wide variety of maladaptive coping behaviors designed to cause weight reduction. Most A-Bs are convinced that gaining weight or being a non-"magical weight" is the source of all their problems. Therefore losing weight or maintaining the magical weight becomes the most important activity in their lives. This belief is held on to tenaciously, even when logical arguments to the contrary are presented.

Since their behavior is totally motivated and reinforced by fear of weight gain, they may incessantly exercise, starve, count calories, abuse laxatives, and structure their entire lives around eating the "proper" foods. A-Bs are overly concerned with their appearance and other people's opinions of them. Their perceived self-worth is dependent mainly on being thin and attractive and they have little internal resources to secure a stable sense of self.

Trying to convince A-Bs through rational discussion to change their attitudes and behaviors is like trying to convince a phobic that the phobic stimulus and resulting anxiety will not hurt them. Even when they agree that their behavior is dangerous and their attitudes are irrational, the negative emotions driving and reinforcing their maladaptive behavior overcome their willpower. A-Bs are not devoid of rational thinking in other areas; their irrational thinking usually only concerns their food, appearance, and body weight.

If the A-B personality were to be classified, one reasonable label would be OCD with dissociative features. The body image distortions are often so pronounced that clients will insist that they are obese and even 'see' themselves as obese when they are starving to death. While there have been many eloquent descriptions of A-B behavior, many authors do not recognize the dissociative (trance) behavior and conversion reactions that are experienced by these clients. Eloquent descriptions add nothing to help A-Bs to change. In fact, much of the literature describing individuals' struggles with A-B often unintentionally romanticize the disorder and make the sufferers feel special.

Bulimics experience the identical emotions and conversion reactions as anorexics, but lose their eating inhibitions and enter trance states during which they binge-eat and vomit. Binge-eating and vomiting may be repeated for hours and even whole days. As with compulsives, an overwhelming urge precedes their binges. A-Bs may be hungry all the time but deny it. Their denial of basic physiologic drives, which includes hunger and sex and other basic emotions, is often a central etiological and maintaining variable for their maladaptive behavior and thinking. Many chronic A-Bs vacillate between starving and binge-eating.

A-Bs are compulsive and perfectionistic, yet shy and timid in interpersonal relationships. However, since A-B is not a disease but a set of learned cognitions and behavior, individual variability is great. All A-Bs make a conscious decision to lose or maintain their weight by restricting caloric intake. There is much data concerning familial risk. However, information such as that bulimics on average have an obese mother is correlational and offers absolutely nothing to facilitate treatment, as no causal variables can be elucidated. So what if bulimics on average have an overweight mother? Many do not. These factors offer nothing to facilitate treatment.

Although there are similarities between anorexics and bulimics, there are enough dissimilarities for them to be classified as different disorders (Garfinkel, Moldofsky, & Gaines, 1980; Casper, Eckert, Halmi, Goldberg, & Davis, 1980). Much of this classification is simply nitpicking as, again, neither anorexia nor bulimia is a disease. The diagnostic categories are simply loose descriptions; any one client may possess a mix of behaviors and traits indicative of both classifications. In fact, categorizing learned behavioral disorders as diseases adds nothing to the treatment. All it does is allow a privileged group of Mental Health Professionals (MHPs), through their lobbying efforts, to get paid by insurance companies.

Bulimics tend more so than anorexics to employ bingeing and vomiting, laxative abuse, and alcohol and drug abuse as methods of avoiding their fears of weight gain and other emotions. Bulimics are generally more obese before the onset of their problem, usually older, and have a family history of obesity (Garfinkel, et al., 1980). Bulimics who binge daily show elevated MMPI scores for schizophrenia, psychopathic deviation, paranoia and psychasthenia, indicating that possibly bulimics are struggling more consciously with their problems (Casper, et al., 1980). Bulimics also have a greater tendency to steal. Obviously bulimics have less impulse control, and this lack of impulse control often results from higher anxiety and subjective distress experienced by bulimics.

My conclusion is that bulimics are actually anorexics who are unable to dissociate emotions as well as anorexics and, therefore, lose control. Supporting this is that chronic anorexics often exhibit the classic *la belle indifference* that characterizes clients who employ dissociation as a means of coping with anxiety. Bulimics more often feel hunger (Casper, et al., 1980), are more extroverted, openly hostile, sexually active, and more easily form interpersonal relationships (Garfinkel, et al., 1980). Also, bulimics more often attempt suicide, supporting that they feel more negative emotions than anorexics and are less skilled at dissociating these emotions.

All of my clients have been chronic A-Bs who have failed in treatment and have experienced their disorder for over 24 months. Bulimics on the average are better candidates for my therapy approach because they are more willing to experience and extinguish the negative emotions that mediate their maladaptive behavior. The consensus of opinion is that chronic A-Bs are resistant to traditional psychotherapy (Halmi, Brodlund & Loney, 1973; Garfinkel, Moldofsky & Gaines, 1977; Morgan, 1975).

Palazzoli (1978) and Minuchin and Roman and Baker (1978) found that family therapy is an effective treatment for pre-adolescent and adolescent A-Bs. Family therapy termed *The Moudsley Model* (LeGrange, et al., 2005) has been shown to be effective for anorexics who develop the disorder before age 19 and had the disorder for less than 3 years. The same approach has been less effective for bulimia (Dare & Eister, 2002). Wilson (et al., 2007) concludes that this approach provides little benefit for patients who have a long history of anorexia. Again, all my clients fit this latter category.

The treatment of choice for bulimia is cognitive behavior therapy (CBT) (Wilson & Fairburn, 1993, 1998). However these clients are seldom entirely free of their problems (Lundgren et al., 2004). Binge-eating and purging is eliminated in 30-50% of the cases (Wilson et al., 2007). CBT, over a period of 1-2 years, has been used to treat anorexics. Pike (et al., 2003) treated 33 anorexics who had been hospitalized and then discharged.; after 50 sessions, only 17% showed full recovery. Nutritional counseling was completely ineffective (Butcher, Mineka & Hooley, 2010).

The question remains: is there a reasonable treatment for chronic A-Bs who are no longer within their family or who need to eliminate their dangerous behavior quickly? The treatment approach described here was developed with chronic A-Bs who were motivated, i.e., who express at least overtly a desire to change.

There is no disease entity or pathogen that causes their maladaptive behavior. No studies have shown that a genetic component causes a person to develop a particular obsession with thinness or any other obsession. One can be predisposed to anxiety, but how this anxiety becomes associated with certain stimuli is explained by well-understood learning principles. People inherit predispositions to temperament characteristics and learning determines how these are expressed in overt and emotional behavior. There is no need to elaborate further concerning hereditary factors in anxiety disorders and eating disorders as it has no relevance to treatment. Maladapative learning can be reversed with proper methods.

MHPs for 80 years have tried to demonstrate a causal relationship between factors like family types with all types of maladaptive behaviors. Identical twin studies over the past 60 years involving twins adopted at birth and reared in different homes have shown that all types of mental disorders and temperament characteristics have a strong genetic base. Thousands of studies that have not controlled for similar genetics between the parents and offspring are meaningless and, adding injury, unfortunately have spawned worthless educational and child-rearing methods (Harris, 2009).

There is strong evidence that A-B, ADs and all types of depression have strong genetic components (Balik & Tazzi, 2004; Fairburn & Hansen, 2003). Genes influence transmitter levels and the functioning of every aspect of the nervous system. The physiological aspects of mental disorders is the domain of medicine and only a competent M.D. should treat these aspects. MHPs definitely do not have this training and few are trained even in the science of behavior. Behavioral science and the science of abnormal behavior should be their focus.

A well-trained behavioral scientist has a lot to offer even though genetic predispositions are part of the problem with clients suffering from mental health disorders. People can extinguish negative emotional reactions, overt behaviors and thinking through the proper application of learning principles. Just because someone is genetically predisposed to react anxiously does not mean they cannot learn to control their anxieties. A-B and ADs are not diseases but groups of learned maladaptive behaviors that are maintained by secondary gains and avoidance behavior. The behavior is the result of early learning and being reared in a society where they have accepted the idea that their self-worth is determined by being thin and attractive. There is no disease pathogen that determines whether or not a person acquires this maladaptive mindset.

Long-term follow-up of treated clients is necessary in order to validate the effectiveness of any treatment, whether it is medical or psychological in nature. My follow-ups ranged from 6 months to 35 years. Treatment time for clients from out of state range from two to 21 days. Clients who live within a reasonable distance are seen initially for 1-2 full days, and then once/week for 2-4 hour sessions. The usual time to treat these clients ranges from 2 to 10 months. My approach was designed for motivated chronic adult cases who failed in previous treatment.

A. Respondent, vicarious, and operant conditioning

Emotional states, urges, compulsions, obsessions, fears, anxieties are all closely related in that they are classically (respondently) conditionable (Pavlov, 1927; Miller & Dollard, 1950). Throughout the history of psychotherapy, MHPs have recognized that anxiety often plays a central role in abnormal behavior and in particular anxiety disorders (Freud, 1936; Wolpe, 1958). In view of the fact that any emotional response such as anxiety or fear can be conditioned to occur to a variety of stimuli makes it reasonable that treatments based on the classical Pavlovian conditioning model should form a major part of any treatment to treat ADs.

In 1920, Watson and Rayner conditioned and reversed a phobia in a 19-month-old child through classical conditioning. In 1940, O.H. Mower combined classical conditioning and operant conditioning to show that conditioned emotional responses (CERs) can act as mediators and reinforcers for avoidance behavior (Mower, 1960). Mower's 2-process theory of avoidance learning is the foundation for the modern successful treatment of OCD (Rochman & Hodgson, 1980; Rochman & Shafran, 1998).

Fear can be learned quickly and serves as a drive or motivator for learning new habits. Anxiety and fear are characterized by the same subjective feelings and physiological correlates. Anxiety is a maladaptive or irrational fear in which the fear stimulus does not signal a dangerous situation or may be vague or unknown. CERs, both pleasant and unpleasant, act as drives and can mediate, reinforce or punish maladaptive and adaptive behavior.

Animal research has shown that as mammals become skilled at performing a coping response that allows them to avoid a noxious stimulus situation, the avoidance behavior becomes automatic and the animal relaxes while performing that behavior. This makes perfect sense, because if an animal stayed in a state of heightened emotional arousal, it would become worn down physically and die. The deleterious effects of chronic stress have been well documented since Hans Selye's pioneering research in the 1950s (Selye, 1956a; 1976a). For example, most of us have learned to fear fire. However, a fire in a fireplace can also elicit a relaxation response. We only become fearful if we get close to the fire and experience an uncomfortable level of heat. Therefore, living in a house with a fire burning in a fireplace produces an adaptable relaxation response instead of a chronic state of anxiety. Obviously a stimulus can acquire anxiety eliciting or relaxation eliciting properties.

Once an avoidance response is learned it becomes resistant to extinction. As the phylogenetic scale is ascended, automatic avoidance behavior becomes more persistent (Solomon, Kamin, & Winne, 1953; Miller, 1952). The role of CERs in the acquisition and maintenance of neurotic, fixated or compulsive behavior in rats has been described by Maier (1949) and later investigators (Feldman & Green, 1967).

In the typical experiment, a rat is placed on a platform (Lashley Jumping Stand) and is goaded by an air puff into jumping off the stand into one of two windows. If the rat chooses correctly, the window opens and it passes through and receives a reward such as food. When the rat hits the wrong window, it does not open, causing the rat to hit the window and fall into a net.

Maier, et al., exposed rats to an unsolvable discrimination task by structuring the situation so that half of the responses were randomly reinforced and half were punished. The rat was unable to learn the correct response because there was no consistent relationship between its behavior and avoiding the punishment of hitting the closed window. This is a frustrating experience for the rats, as it would be for humans. The situation reminds one of depressed clients who have said, "No matter what I do, it's wrong."

Rats experiencing the above situation become fixated in their responses and persistently jump to one window without altering their choice. Maier termed this stereotyped behavior as "fixated behavior." Fixated behavior is defined as a "persistent mode of behavior which has outlived its usefulness or has become inappropriate" (Chapin, 1974). Maier, et al., then exposed the fixated rats to a solvable discrimination task, and found that the majority of the rats persisted in their fixated behavior. A rat who had become fixated in jumping left persisted in jumping left despite being shown that the right window was open. When the rat made an incorrect/fixated jump, it positioned its body in midair in such a way as to minimize injury. Often this position made it impossible for the rat to get through the window even if were to make a correct choice. Anthropomorphizing, the rats expected that their responses would not be successful and were unable to alter them.

Fixated behavior is not alterable by simple reinforcement procedures because the rat never performs a correct response that can be rewarded. Punishing the incorrect responses may only increase the fixated behavior. By increasing anxiety it appears that animals revert to previously overlearned coping behavior and abandon recently learned and more adaptable behavior. This strategy is used in sports by making one's opponent emotionally upset so they abandon their skilled reactions and revert to unskilled and more reflexive primitive behavior. This is called getting an opponent "out of his game" or "psyching him out."

Anthropomorphizing again, the rat acts as if he knows which is the right window but cannot alter his fixated-compulsive ritualistic behavior. The similarity between experimentally produced fixated behavior in mammals and obsessive-compulsive behavior in humans is striking. A more detailed model of obsessive-compulsive behavior in humans is presented in Chapter L.

Classical conditioning works best on autonomic nervous system responses (smooth muscles and glands) that are influenced by emotional states. Throughout the past 80 years, all varieties of emotional states have been shown to be elicitable through conditioning by almost any stimulus that is perceptible or near the limen of conscious perceptibility. Stimuli from all sensory modalities can serve as conditioned stimuli (CS), and through association can acquire the ability to elicit simple and complex emotional responses. Of course some CS-CRs (conditioned responses) associations are more easily learned than others because they have more survival value. In other words, how we react emotionally in any situation including cognitive thinking situations is learned and can be extinguished or unlearned.

Of course most of our anxiety responses have not been directly conditioned by experiencing noxious stimuli. It has been conclusively shown in primates and humans that neutral stimuli can acquire the property of eliciting strong anxiety by mammals observing anxiety reactions in others (Bandura & Rosenthal, 1966). This type of learning is termed *observational* or *vicarious learning*.

Fears (anxieties), being learned responses, are alterable by learning procedures. One proven way to eliminate unrealistic or irrational fears (fears elicited by stimuli not usually followed by dangerous situations) is to expose the organism to the stimuli that elicit the anxiety and prevent it from being reinforced by an aversive event. In this way, avoidance of the anxiety is prevented and extinction can occur. Repeatedly presenting any stimulus to an organism that has acquired an anxiety response and not allowing the organism to avoid that anxiety stimulus and experience the anxiety until the anxiety is gone is termed *flooding*.

Counter-conditioning has also been shown to decrease anxiety responses. Counter-conditioning involves pairing the anxiety-eliciting stimulus with a stimulus that elicits an incompatible response such as relaxation. The basic idea is that relaxation elicits parasympathetic activity which is incompatible with the sympathetic activity that accompanies anxiety. Wolpe (1958) terms this *reciprocal inhibition*, and forms the basis of systematic desensitization. Wolpe's procedure involves successively presenting parts of the anxiety-eliciting stimulus along with having the client relax. The anxiety is reciprocally inhibited in small increments. This method has been shown to be effective for the treatment of phobias (Bandura, 1969; Wolpe, et al.)

In spite of systematic desensitization's practical use, it is still controversial as to what the therapeutic ingredients are. Yates (1970) concluded that extinction is the important variable. Research has shown that forming the stimulus hierarchy by starting with a dissimilar stimulus to the original anxiety stimulus and successfully using more similar stimuli does not facilitate the elimination of anxiety.

Flooding (exposing clients to as much of the anxiety stimulus as they can endure) probably works better than systematic desensitization on phobias. *Extinction* (the non-reinforced elicitation of anxiety) is the major variable in techniques such as flooding and systematic desensitization. There have been valid criticisms of classical conditioning theory, especially on the grounds that it ignores cognitive variables. Expectations and cognitive variables are potent variables in behavior. Bandura (1977) offers an alternative explanation for why procedures such as systematic desensitization and flooding produce behavioral change. His theory is termed *Self-Efficacy Theory*.

Self-Efficacy Theory proposes that desensitization, flooding, and similar techniques which involve clients exposing themselves to the anxiety eliciting stimuli work because they increase clients' efficacy expectations (the subjective estimate of one's ability to cope with threatening situations). The therapeutic ingredient according to Bandura is that clients have internalized a heightened sense of control over their lives (Perlmutter & Monty, 1980). Bandura's alternative explanation is congruent with Seligman's (1975) concept of learned helplessness described later.

There is validity to both the conditioning and the self-efficacy theories, and they complement each other. Both predict that exposure to anxiety-eliciting situations will cause decreases in clients' reports of fear and avoidance behaviors. Bandura's theory offers an explanation, although unproven, of why exposure techniques occasionally fail. According to his theory, exposure techniques will not cause extinction of anxiety and avoidance behavior unless the client internalizes the cognition or attitude of self-efficacy.

Another variable that can block extinction is that some clients who are gifted in the use of dissociation can effectively ignore the negative emotions during flooding or exposure experiences. This dissociation is more thoroughly explained in Chapter J. Both cognitive and conditioning aspects of A-B and AD clients must be considered for lasting improvement to occur. Altering expectations is important, and one of the most powerful and most neglected tools for doing so is hypnosis.

A basic law of hypnosis states, "All behavior follows expectations, and as people expect to behave, they tend to behave." I had a client who was using my hypnosis tapes very effectively for weight reduction. While he used the tapes, he was steadily losing weight. A year later, I asked how he was doing; he stated he had stopped using the tapes because he heard on TV that subliminal messages do not work. Clearly he was confused as to the difference between hypnotic suggestions and subliminal messages, and this confusion destroyed his expectation that hypnosis would work. Hypnosis and its various laws are explained in Chapters J, K, L, N, and O.

Operant responses (emitted or voluntary responses mediated by the somatic nervous system) increase in strength and frequency when followed by reinforcement, which can be an increase in pleasure or a decrease in an unpleasant event. When reinforcement is withheld, the organism's operant rate of responding decreases (extinguishes). Punishment (following an operant response by an aversive stimulus) decreases the response strength and frequency, although the effects may be variable depending on the type of response being punished. Although there are controversies concerning the role of reinforcement and punishment in operant learning, behavior is definitely influenced by its consequences. Covert processes (thinking, cognitions, or beliefs) often manifest properties similar to operant responses and are influenced by reinforcers that follow their emitants

(Homme, 1965). The application of learning principles to alter thoughts and cognitions is termed 'cognitive behavior modification' or 'covert conditioning' and is an active clinical research area. There is some controversy caused by the lack of control over clients' increased expectations that the techniques will aid them. So again, potentially powerful cognitive variables in human behavior are encountered: expectations and attitudes.

All therapists use the above procedures, whether or not they realize it. MHPs' unsystematic use of cognitive and behavioral procedures explains why non-directive, insight-oriented therapies help people with ADs better than control subjects who receive no therapy. This also explains why minimally trained MHPs achieve the same rate of success as the most highly trained clinical psychologists who use non-directive techniques (Dawes, 1994). My conclusion is that the vast majority of MHPs who use non-directive techniques are equally effective or ineffective in the treatment of A-B and ADs.

B. Two-Process Theory of Avoidance Learning

The idea that CERs and emotional states in general serve as mediators and/or reinforcers for operant responses forms the basis of Mower's (1947) two-process theory of avoidance learning. For example, someone who repeatedly experiences anxiety in a particular situation will learn to experience anxiety in that situation on subsequent occasions. The anxiety mediates voluntary (operant) behavior designed to avoid that situation. The reduction of the anxiety when the situation is successfully avoided reinforces those successful or partially successful avoidance responses (Mower, et al.; Resacorla & Solomon, 1957).

Consider the case of a school phobic who developed his fear of school by experiencing punishment in school. By the contiguous pairing of school and anxiety-eliciting events, even the thought of school elicits anxiety. Following the negative emotional experience in school, the child feels more anxious the closer he gets to school. The closer he gets to school, the more the original stimuli that had been paired with the negative emotions are approximated, so the more anxious the child becomes. The anxiety becomes so great that it motivates him to make the operant response of turning and running away. The farther he is away from school, the more anxiety is reduced, reinforcing the avoidance response of running away.

The treatment methods based on the dual-process theory in combination with hypnotic (cognitive) principles is the best model for explaining the origin and treatment of A-B, OCDs and dissociative disorders. Hypnotic principles bring in the role of expectations and attitudes towards the efficacy of treatment and other variables such as clients' self-images and other attitudes that have to be taken into consideration.

It is reasonable that once successful avoidance is learned, the anxiety of that situation should decrease and the avoidance behavior should become automatic. It would be physiologically disadvantageous for an organism who encounters daily dangerous situations and has learned to avoid them to successfully stay in a continual state of fight or flight. A continual state of stress is extremely debilitating (Selye, 1956 & 1976).

This does not mean that people would not experience anxiety when in close proximity to the anxiety-eliciting situation. People do not forget that that particular situation was paired with an unpleasant stimulus. If people did, they would not continue to successfully avoid the noxious situation. The memory of the relationship between the situation and the noxious event must still be present in order for people to maintain their avoidance and survive.

Importantly the active avoidance of noxious stimulus, whether overt or covert, allows the negative stimulus-emotional response association and negative expectations to remain. The avoidance behavior and expectation of a noxious experience, whether it is adaptive or maladaptive, becomes reinforced by the reduction of negative emotions. Cognitive avoidance strategies, such as switching attention to other stimuli, are also employed to avoid the negative emotions. As avoidance is rehearsed, the avoidance behavior becomes more automatic or 'unconscious.'

Hersen and Detres' (1980) treatment of bulimia by flooding and response prevention (FRP), further discussed in Chapter E, is based upon the idea that bulimia is a means of reducing tension. Their analysis is in agreement with my theoretical account of the origins and successful treatment of A-B and OCD. It has been my experience that if an urge, panic, or emotion precedes and mediates a compulsive act, then the two-process theory applies. However, there may be maladaptive compulsive behaviors that are automatic or maintained by expectancies and are no longer mediated by CERs. These behaviors respond best to an operant conditioning therapy involving the punishment of the maladaptive behavior and the reinforcement of an adaptive and incompatible response. Implanting positive expectations through hypnosis in addition to imagery conditioning often reduces compulsive behavior patterns.

An example of an automatic behavior is when a smoker reaches for a cigarette without realizing it. As smokers often smoke to decrease negative CERs and to induce relaxation, smoking becomes a secondary reinforcer by being repeatedly paired with relaxation. It is this type of smoker who is resistant to the simple operant behavior modification techniques. The same concept may apply to clients exhibiting OCD. These clients may respond best to first extinguishing the respondently conditioned urges (CER) that mediate and reinforce the active smoking. Following extinction, aversive conditioning can be used to punish the compulsive behavior. An incompatible behavior can be substituted for the maladaptive behavior and must be reinforced.

For example, behavior incompatible with smoking can be cued by the thought that precedes smoking. Clients are instructed and also given the post-hypnotic suggestion to emit an incompatible and relaxing behavior to substitute for smoking. The suggested behavior is usually to take a deep breath, hold it, and then relax with a pleasant image while exhaling. I have found significant success using this approach with a large number of cigarette smokers.

C. Extinction, Spontaneous Recovery, and the Unconscious Mind

Throughout the last hundred years, therapists have stated that the unconscious mind has no sense of time. I view the unconscious mind as that part that contains CER associations that the client is successfully avoiding. The successful avoidance of the eliciting stimuli (CS) never allows the CERs to occur and undergo extinction. In order for extinction to occur, the CERs have to be elicited and not followed by an unpleasant event. Therefore CERs appear to be independent of time decay and are eliminated only through extinction. Many MHPs have remarked that the unconscious mind has no sense of time.

This has adaptive value. Consider the example of a child who witnesses her friend being bitten by a poisonous snake and convulsing to death. If she does not see a snake for 50 years, she will still react automatically with anxiety upon viewing one. CERs do not decay over time. Childhood traumas and their associated CERs do not spontaneously decay, but they can be extinguished. CERs can be acquired in seconds and extinguished in a few hours or less.

Insight into how and when the CERs were acquired does nothing to help extinguish them. Again this makes adaptable sense. As an example, if one could just will away these negative CERs so that they would no longer be fear-eliciting and signal danger, one might pick up a poisonous snake, be poisoned, and die.

Extinction is an active process in which the CER is inhibited as opposed to a gradual loss of the CS-CER association. Spontaneous recovery, first demonstrated by Pavlov (1927, 1960), is produced by extinguishing a CR and waiting for a time period, and then presenting the CS again. The extinguished response will usually return but at a weaker strength, indicating that the extinction process decayed. Spontaneous recovery substantiates that extinction is an active inhibitory process superimposed over the CER and has a decay time.

Spontaneous recovery is more the rule than the exception when a CR is not over-extinguished. Over-extinction occurs when extinction trials are continued after the CR amplitude is zero. This usually produces lasting extinction, although occasionally even it may need to be performed again. Therefore, it should not surprise a therapist when a client experiences the spontaneous recovery of previously extinguished maladaptive behavior and CERs. Extinction again must be carried out.

Respondent extinction is an important variable in psychotherapy and explains the success of behavioral techniques such as flooding, FRP, and systematic desensitization (Bandura 1969). Wolpe argues for reciprocal inhibition being the therapeutic ingredient in systematic desensitization and, as stated, Bandura offers evidence that increased self-efficacy is the therapeutic variable (Bandura, 1980; Bandura, Adams, Hardy, & Howells, 1980). In any case, clinically these concepts are compatible and can be combined. If relaxation, reciprocal inhibition and self-efficacy are not shown to be necessary for the extinction of negative CERs, they certainly facilitate extinction.

D. Abreactive Extinction and Hypnotherapy

Eliciting abreactions (catharses) under hypnosis for the treatment of hysterical conversion reactions began with Breuer and Freud in the early 1900s and formed the basis of Freud's psychoanalytic conceptualization of unconscious motivation. Abreaction is defined as an emotional release or discharge resulting from recalling a traumatic experience that had been repressed because it was consciously intolerable (Frazier, Campbell, Marshall, & Werner, 1975). The cathartic method utilized abreactions and involved the release of ideas through talking about conscious or unconscious emotionally charged material (Frazier, et al.). Breuer's cathartic method was a failure. Hospital records discovered by Ellenberger (1972) confirmed that the classic case of Anna O. was a failure. She continued to rely on morphine to ease the hysterical problems that Breuer was supposed to have removed by having her ventilate her emotions through the cathartic method.

The failure of the cathartic method is explained by the misconception that insight into the origins and reasons for the symptoms and the release of energy that the client experiences during the abreaction were the therapeutic agents. Psychoanalysts abandoned the cathartic method and opted for the invalid analytic approach. At the same time Watson and Raynor's (1920) classic demonstration of the acquisition and extinction of a phobia in a human was ignored by psychodynamic theorists. Watson & Raynor's, et al., work was well known and reviewed in every introductory psychology textbook since the 1930s. However, the vast majority of MHPs ignored their work and followed the misguided theories and techniques expounded by Freud. Freud was a cocaine addict for a number of years and later admitted that this drug use had clouded his mind.

Pavlov's 1927 pioneering work, which formed the basis of Watson & Raynor's classic demonstration, was published in English in 1927. Pavlov clearly illustrated the acquisition, extinction, spontaneous recovery, and generalization of conditioned responses. Sadly, the majority of MHPs ignored his work as well and also opted to follow the ridiculous Freudian methods.

When one analyzes the abreactive process in terms of extinction and spontaneous recovery, an explanation for Freud's and Breuer's failures becomes obvious (Bandura, 1969). As stated, the most reliable way to eliminate CERs is through extinction. Once a CER is elicited by a CS and no noxious (unconditioned) stimulus follows, the CER diminishes after successive presentations of the CS alone. The autonomic correlates to the extinguishing CER diminish along with the extinction procedure, although this may not exactly parallel the extinction curve of the client's subjective feelings (Lifshitz & Blair, 1960).

A repressed thought, if there is such a thing, may be conceptualized as a CS that the subject has successfully learned to avoid by making a variety of responses such as fidgeting, nail-biting, cognitively ruminating (Bandura, 1969; Miller, 1975) or through dissociation (Keppner & Meehan, 1982). These successful avoidance strategies maintain the unadaptable behavior because the CER mediator is never elicited and allowed to extinguish. The maladaptive avoidance behavior also becomes more persistent through rehearsal.

The mediating CER and/or cognitive state is usually triggered by both overt and covert conditioned stimuli. When these stimuli are encountered, the client feels the beginning sensations of the CER. As soon as clients feel these sensations they rapidly and often automatically (unconsciously) begin various adaptive and maladaptive coping maneuvers. The quicker and more successful these coping behaviors are, the less the probability that the CER will be even partially elicited. Therefore, the negative mediating emotions are guarded from extinction. If the behavior is starving or binging and vomiting, the consequences of temporarily avoiding the mediating emotions through these maladaptive behaviors may become life threatening. The more temporarily effective the maladaptive behavior or coping responses are in guarding against the negative CERs, the more resistant the avoidance behavior is to change and the more frequently it may occur.

The relaxing or avoidance behavior initially causes a reduction in tension and is reinforced. However, if the avoidance behavior also elicits negative emotions, the client will again avoid these emotions by the same avoidance behavior. As clients attempt to force away urges, negative emotions, thoughts, and fears of loss of control, tension increases. Using one's willpower to force out of consciousness unwanted thoughts has repeatedly been shown to cause these thoughts to re-emerge with greater intensity (Wegner, 1987; Wegner & Erskine, 2003).

In the pioneering work *On Thought Suppression*, Wegner et al., instructed subjects to suppress thoughts of a white bear, and found that willful suppression caused an increase in thoughts of white bears. The thoughts can occur with increased intensity during the suppression or after the suppression. This forms the basis of the ancient law of hypnosis, termed the *Law of Reverse Effect*: the harder one tries to will one's self to not think of something, or to suppress an emotion, the stronger it becomes. The harder people try not to feel an emotion, the more they will feel it. This increases the negative mediating emotions which in turn cause more maladaptive coping. A vicious positive feedback loop is formed, locking the client into a cycle of repeatedly performing the compulsion. The clinical application of this law is explained in Chapter L: "Hypnobehavioral Model of Obsessive-Compulsive Behavior and the Unconscious."

For example, compulsive binge-eating and vomiting begin with the client experiencing negative emotions or hunger which causes her to seek food. Food is inherently reinforcing and is a conditioned relaxer (reinforcer) by having been associated with pleasant events. She eats, experiences relief, thus reinforcing eating as a means of tension reduction. However, following the reduction of negative emotions, the knowledge of having eaten elicits the fear of weight gain. This powerful fear is reduced by vomiting; this in turn means the food has then been wasted and she has lost control. This triggers negative emotions such as guilt and fear that the client again attempts to reduce by eating. She therefore becomes locked into the cycle of repeatedly binge-eating and vomiting. The cognitive aspects of this behavior pattern are discussed in Chapter M: "The Bulimic Cognitive-Behavioral Feedback Loop."

The basis for compulsive binge-eating and vomiting occurs when the consummatory behavior of eating takes on the properties of anxiety reduction and then anxiety elicitation. This pattern is described in the aforementioned Chapter L and is related to trance behavior in Chapter J: *Anxiety, Dissociation, and Conversion Features of Anorexia-Bulimia*.

Various CSs that elicit the negative CERs are encountered in a wider range of situations which mediate the avoidance of those situations. Many A-Bs avoid social situations involving food or any situation where they eat and cannot find a way to vomit. The avoidance of situations that were pleasurable or could be positively reinforcing restrict clients' lifestyles and range of positive reinforcers that maintain their level of enjoyment. As clients avoid enjoyable situations and activities, they become inactive, depressed, and helpless because these activities that made their life worthwhile are no longer available. This is the beginning of the internalization of the attitude of helplessness that often accompanies depression, OCDs, ADs, and A-Bs, and helps explain why 68% of anorexics and 63% of bulimics are depressed (Hudson, et al., 2007). The cognitive behavioral model for treating depression is explained in Chapter H: *Cognitive-Behavioral Treatment of Depression*.

While in hypnosis, clients can be guided into imagining or 'regressing' to CERs (traumas), thoughts, beliefs, or ideas that maintain their problems. The urges that addictive compulsives and A-Bs experience can be elicited by having them imagine the urge or regress to the mediating emotions and thoughts that underlie their maladaptive behavior, or by imagining eliciting stimulus situations. Similar to abreactive extinction, when clients imagine or experience a CS that elicits urges etc., they begin to experience those urges and their autonomic correlates. If the CERs are triggered rapidly, the anxiety may flood the client and cause an emotional upheaval (abreaction). Since no noxious stimulus is present to reinforce the CS-urge association, the CERs underlying the urge begin to extinguish.

The elicitation and extinction by hypnotic regression, hypnotic imagery, and in-vivo extinction of the urges that precede compulsive binge-eating and vomiting forms an important part of my treatment. The extinction of such urges by using the aforementioned methods and not allowing the clients to reduce the urges by any other means has been applied in the treatment of drug abuse.

Gotestam and Melin (1974) had four amphetamine addicts visualize scenes of giving themselves injections. The imagined scenes elicited urges and feelings associated with injecting the drug. Throughout repeated extinction trials, the autonomic correlates (pulse rate, respiration, and galvanic skin response) declined as extinction proceeded. Three of the four clients reported no drug use on a four month follow-up.

In-vivo extinction with narcotic addicts had been reported as early as 1931 by Rubenstein. Wikler (1973) hypothesized that heroin cravings are CERs that can be extinguished by eliciting them and not allowing them to be reduced or avoided. O'Brien (1974) tested Wikler's hypothesis using heroin addicts who received naloxone (which quickly blocks the narcotic effect of heroin, thus preventing the addicts from experiencing euphoria or any other reinforcing effects).

Six clients administered a total of 50 self-injections of heroin following the administration of naloxone. The clients reported that after about 15 trials, the feelings associated with the injections gradually moved from euphoria upon injection to experiencing it as aversive. Four of the six clients reported that they were drug-free on an eight month follow-up. Other researchers reported that the clients using this technique must still take part in a methadone or traditional drug treatment program

(Hurseler, Gewirtz, & Klebr, 1976).

It is not my intention to enter this controversy. However, the aforementioned studies suggest that urges, at least the ones that narcotic addicts experience, are CERs and become associated with and elicitable by a wide range of stimulus situations. O'Brien (1974), Himmelsback (1942), and Winkler (1973) have shown that the powerful physiological effects and cravings, which heroin addicts experience when presented with stimuli associated with "shooting up," are not elicited in a drug-free environment such as jail. Therefore, these cravings do not extinguish while addicts are in these environments because they have not undergone non-reinforced elicitation. Moreover, these cravings, because they have not been extinguished, will be triggered years later when the addict returns to the drug-taking environment. Extinction in the new environment must take place.

The same phenomena have been demonstrated in rats. Wikler & Pescor (1970) trained rats to self-administer morphine. After the rats were removed from the experimental situation for a year, they returned to self-administering the drug when returned to the original drug-taking environment. Lietenberg, Rawson & Bath (1970) demonstrated that it is critically important for such behavior to be non-rewarded in order for the behavior to be extinguished.

In the late 1980s, I treated a cocaine addict. She was a law-enforcement officer who enjoyed her job and was happily married. Importantly, she was gaining no secondary gains from her drug use and exhibited a tough character. Additionally, her drug use was not connected to social events; she hid her addiction from others and would inject only when alone. While many drug abusers obtain a perverted sense of pride from being an addict, this client experienced only shame and anger about her behavior.

I instructed her to rehearse all her normal actions preceding injecting while in the actual room where she usually injected and with the cocaine present, and to orally record all of her actions, emotions, and urges. She was to also work at trying to increase her craving for the drug and to become as miserable as possible. Her attitude was one of aggression toward her problem, and she was able to do this successfully. She reported experiencing horrible cravings, but continued flooding herself in the presence of the drug until the cravings were gone. Whenever she experienced a craving, she again practiced the flooding exercise.

After two weeks of daily practicing, she was elated with her success at extinguishing her addiction. On a two year follow-up, she was still drug-free. Immediately after her success, she wanted to help others and in particular her brother, whose life was rendered useless by drug addiction. From her description of her brother, he appeared to have neither the same desire to stop nor the toughness of character she had. I told her to try it, but explained the difference between her and her brother in terms of their character, lifestyles, and environments. Her efforts with her brother failed as he was not willing to follow through with the extinction processes.

A well-known psychiatrist once stated he had a 100% cure rate with his eating disordered clients. His evidence: none of them returned after treatment -- although no one ever followed up on any of his patients. Using this standard, I can state I have a 100% cure rate with alcoholics, because if an alcoholic was not willing to use the prescription medication Antabuse, I rejected him. Over a 20-year span, I treated one successfully and rejected approximately 100 others for that reason. Similarly, unless clients with A-B or AD are willing to undergo FRP and other exposure techniques, I will not treat them.

There are similarities between narcotic addicts, alcoholics, bulimics and OCD clients. All experience overwhelming urges and lack volitional control when they encounter certain stimulus situations. The urges that bulimics experience when encountering binge-eliciting stimuli parallel the urges that OCDs, drug addicts and alcoholics experience. Salzmann (1980) classifies A-B as an addictive-compulsive behavior, and I concur.

Not only can external stimuli acquire strong urge-eliciting properties, but so can internal (covert) stimuli such as specific emotional states, general uneasiness, and a wide variety of cognitions. Most any stimulus can acquire the property of eliciting strong urges to commit maladaptive acts by simply being contiguously paired with that urge. Any behavior having been paired with the reduction of an urge can also acquire positive reinforcing properties.

The similarity of extinction curves for urges that bulimics and heroin addicts experience support the hypothesis that these urges are CERs and, therefore, extinguishable. Confirmation of the respondent conditioning model is provided by O'Brien, Testa and O'Brien (1977), who found that heroin addicts generalize their emotional responses elicited by heroin administration to a variety of situations that have been associated with heroin usage. I have observed similar generalization of urges in bulimics. Stimuli which elicit urges to binge and vomit often involve food previously associated with pleasant emotional states. As time passes, the urges and binge-eating and vomiting are experienced in a wider variety of situations and become elicitable by those situations. Therefore, binge-eating and vomiting become a coping method for a variety of emotions. Negative feelings and cognitions act as CS as easily as external stimuli.

The abreactions witnessed by psychoanalysts were simply strong CERs, emerging as the client became conscious of the CS that elicited the troublesome emotions. Usually clients do not completely encounter the eliciting CS and therefore the CERs never completely extinguish or slowly extinguish as parts of the relevant CS are encountered. Slow and partial extinction of troublesome emotions during dynamic therapy occurs through language-mediated elicitations. This explains why analysis and non-behavioral techniques take such a long time to produce results.

Freudians hypothesized that the troublesome emotions were repressed, and this formed a cornerstone of their therapeutic treatment. Repression, or the motivated forgetting and blocking out of consciousness of traumatic memories, is an essential concept in Freudian theory. Repression has been the subject of much controversy and probably does not exist as Freudians define it (Holmes, 1974). It is also not necessary to explain abreactive extinction as more normal cognitive strategies do so quite well. Traumatic experiences obviously are temporarily forgotten, and then clients are reminded of them when the relevant CS are encountered. People do not forget a trauma; they simply have many periods where they do not think about it. When certain stimuli are encountered, they then experience the CER.

Obviously, total repression of memories of traumatic events is unadaptable. Humans throughout history needed information to be rapidly and consciously accessible concerning past traumatic events in order to avoid those events in the future. Memories of traumatic events stay easily accessible in order to protect people. Post-traumatic stress disorder (PTSD) shows that, quite contrary to ideas of repression, traumatic events often keep intruding into conscious awareness.

Freud abandoned the use of hypnotic regression to elicit abreactions in favor of free association and his speculative interpretive procedures because the abreactive process only temporarily inhibited the clients' anxieties. Freud's cathartic method would have proved more beneficial than his unscientific analytical approach had he extinguished the troublesome emotions by repeatedly abreacting them, and then taught the the client adaptive behavior that was incompatible with the maladaptive behavior. However, as mentioned, his thinking had become clouded because of cocaine abuse and became more and more bizarre.

Obviously the abreactive process failed for Freud and Breuer because the emotions were elicited only once. Complete extinction seldom occurs on one trial. Even if extinction did occur with one trial, spontaneous recovery of the extinguished CER would be expected. As stated previously, in order to obtain complete extinction it is necessary to present the anxiety-eliciting stimuli repeatedly without reinforcement (Bandura, 1969).

Usually the CERs' strength temporarily increases during initial extinction trials and then diminishes following repeated elicitations. This explains why therapies that elicit abreactions, such as primal scream therapy, etc., may leave some clients in a more anxious or sensitized state than if the procedure was never used. I encountered a client who had been hospitalized for extreme anxiety following a primal scream experience. The period of "working through," or whatever the post-elicitation period is termed by insight-oriented therapists, is simply time for unsystematic extinction to occur by non-reinforced, language-mediated elicitations.

Word association procedures that have been used to elicit abreactions are consistent with the above interpretation. Words are CS that elicit the avoided mediating negative CERs. Non-reinforced elicitation of the CERs by repeatedly talking about the CS or eliciting words causes those emotions to extinguish. For example, a client who had been sexually molested as a child began to experience an abreaction when certain stimulus words were presented. These words evoked images similar to the denied molestation. Having her imagine scenes similar to the traumatic situation continued to elicit strong negative emotions. Often one image will lead to another, and the order of elicitation is usually from the least anxiety-eliciting image to the most. This process has been described by Miller (1975) as similar to peeling off the layers of an onion. As one layer of emotions is extinguished, another layer that would have been extremely intense but is weakened as a result of the previous extinction will emerge and begin to extinguish.

The use of word association to produce abreactions is facilitated when the client is in a hypnotic state. In the aforementioned case, the client's fears and conversion reactions subsided by repeatedly eliciting the negative feelings that mediated her response of jaw clenching, fear of men, and a variety of CRs such as pain in the genital area. She reported six prior instances of sexual abuse. Whether or not all six of these situations were real or imagined does not matter; it is what has been encoded into the mind as real which matters. The client's fears, insomnia, etc., were eliminated simply by the abreaction and extinction of the emotions subserving those behaviors. As the mediating emotions were extinguished, her maladaptive behaviors also extinguished, including her anorexia.

As the CERs are being elicited, the intensity of the conversion reactions increase and then subside as the mediating emotions become more intense and then extinguish, indicating that they are CERs. The following case study supports the conclusion that emotions occurring during abreactions are CERs and extinguishable through repeated elicitations.

Lifshitz and Blair (1960) demonstrated the progressive decline of emotional responses by repeatedly eliciting abreactions via hypnotically regressing a subject to a relevant traumatic episode. The subject was a 23-year-old female who, through hypnotic regression, revivified a frightening near-drowning incident, which occurred at ten years of age. While under hypnosis, she recalled this episode seven times. Throughout each of the recalls, the autonomic correlates of her anxiety were monitored. Through repeated non-reinforced elicitations, the autonomic indicators extinguished. However, the subject continued to exhibit marked autonomic responses to other unrelated incidents. This indicated that specific negative emotions were extinguishing and ruled out the possibility that a general adaptation or fatiguing of her central nervous system was taking place.

I found with over one hundred clients that having them repeatedly elicit CERs, urges, etc. which precede, mediate and reinforce a compulsive act, and by following this elicitation with relaxation, the urges or CERs extinguish. Extinction can be accomplished without regressing a client to traumatic situations that explain how and when the inappropriate associations and the maladaptive coping behavior were originally learned. The mediating anxiety and its association with food is usually learned vicariously. Again, this anxiety is easily elicited and extinguished by the use of these techniques.

From the standpoint of extinction, it does not matter how the mediating emotions were learned; what does matter is that the CERs must be repeatedly elicited and not avoided so extinction occurs. The elicitations can be made by either presenting the real CS (in-vivo extinction) to clients or by having clients imagine them. For an example of how the combination of flooding (forcing the client to encounter the fear-eliciting stimuli) and response prevention (preventing the client from engaging in a compulsive ritual or maladaptive behavior that would allow avoidance of the emotions) is used for the treatment of compulsive behavior, the reader is referred to Chapter E: *The Behavioral Model for Obsessive-Compulsive Disorders*.

Abreactive extinction has been used to elicit and extinguish traumatic CERs as a part of behavioral treatment. Most modern therapists agree that hypnotically induced abreactions along with hypnosis in general form part of an effective treatment for anxiety disorders (Brown & Fromm, 1986; Peebles & Fisher, 1987; Spiegel, 1993; Nash & Barnier, 2008). Many of the therapeutic gains cited for a variety of therapies are explained by behavioral processes and, in particular, extinction.

My experiences are congruent with Mikulas' (1974) conclusions concerning behavioral variables in psychotherapy. Mikulas (1974) argues that, "The assumption of psychology is that there is a set of laws that describe factors that determine a person's behavior. ... the closer the treatment program comes to utilizing these laws, the more effective it is. It is not known exactly what these basic laws are, but the experimental psychologist believes that the information from the experimental laboratory is the best approximation we have at present." (p. 199)

Returning to the topic of the clinical utility of the abreactive extinction, Bandura (1969) summarizes that, "Abreactive procedures are probably best suited for producing rapid and stable extinction to emotional responses developed in traumatic conditioning situations provided that the threat stimuli are no longer present" (p. 411). "From a learning point of view, the critical therapeutic factor is repeated elicitation of emotional responses without reinforcement rather than an energy discharge or the historical insight" (p. 411). I disagree with Bandura's and others' conclusion that the abreactive procedure is best suited for behavioral problems resulting from a single traumatic episode. Whether or not a single traumatic episode occurred, clients can be made to experience the CERs that mediate maladaptive behaviors or they can be elicited by imagery. In either case, repeated elicitations are required to extinguish the emotions. I do agree with Bandura that the insight clients gain as to the origin of their problems is of little therapeutic benefit. Also, clients' self-efficacy and expectation of success increase following abreactive extinction and facilitates extinction. This is in agreement with Bandura's (1980) formulations concerning the role of the expectation of self-efficacy in extinction. Techniques producing predicted results and relief enhance clients' expectations of success and increase their faith in the therapy.

If Bandura's first statement was correct, the abreactive extinction procedure would have limited therapeutic value. The learning history of most clients show that the negative CERs were reinforced by more than one event and the mediating emotions and operant behaviors are the result of an accumulation of experiences and vicarious learning. Seldom does one traumatic incident trigger a problem. Usually a client is sensitized early as a result of one or more experiences, and even a minor trauma later may trigger the maladaptive overt behavior and anxiety that then brings the client into treatment.

Traumatic events that produce extreme anxiety do not have to involve physical trauma or abuse. Any event that elicits anxiety is traumatic. For example, to an 8-year-old boy, experiencing a broken arm from falling out of a tree would not be as traumatic as his mother telling him to "shove it up your ass" when a sibling breaks a prized belonging and nothing is done about it. Separation anxiety is elicited by this event but not by the breaking of his arm. Separation anxiety may be the base anxiety that originally mediates most maladaptive behavior and cognitions in A-Bs. Bregers' (et al., 1974) and others conclusions concerning separation anxiety are explained later in Chapter J.

William Bryan and later medical hypnoanalysts, such as George Honiotes, adhered to a model of AD based on the medical model for allergies and psychoanalysis. Medical hypnoanalysts state that early events (termed *initial sensitizing events*) set the stage for emotional problems, and later traumatic events (termed *symptom producing events*) trigger the observable maladaptive behavior. These events are identified and analyzed, and the maladaptive behavior is changed through the combined use of hypnotic regression, abreaction, and dream analysis.

My criticism of hypnoanalysis and psychoanalytic procedures is that abreactions should be elicited repeatedly until the mediating negative emotions are extinguished. I had the opportunity to view numerous examples of this procedure being administered in the treatment of a wide variety of ADs by Honiotes, Sanders, and Bonelli at the Midland Institute (Joliet, Illinois) in the 1970s. Although there has been little research to assess the validity of these procedures, many impressive case studies were witnessed involving short-term therapeutic interventions utilizing this approach.

The learning or conditioning model is congruent with the hypnoanalytic model in that both models state that early sensitizing events are negative CERs which sensitize the central nervous system to the acquisition of other and sometimes stronger CERs. Sensitization (increased sensitivity of a neuronal system) is a common phenomena and has been demonstrated in a wide variety of neuronal systems, including the spinal cord separated from the brain (Thompson & Spencer, 1966; Keppner & Roccaforte, 1974), and even in plants (Keppner, 1974).

Medical hypnoanalysts often use word association procedures in order to elicit and extinguish troublesome emotions. Again, these methods are in close agreement with the conditioning model in that words are the CS which elicit the negative CERs that require extinguishing. In this manner, the mediating emotional states and the reinforcers for the maladaptive behavior can be eliminated. Following extinction, the operant behaviors are extinguished or inhibited by post-hypnotic suggestions and by prompting the client to practice new behaviors. Medical hypnoanalysts are aggressive and advocate a client taking an active attitude toward living and coping with their problems. This attitude mediates more adaptive coping behavior that is subsequently reinforced. The medical hypnoanalytic approach and similar approaches are effective because they employ methods which include the systematic and unsystematic use of the basic behavioral and hypnotic principles described in this text.

Analysis is not the therapeutic variable when viewed independently of the behavioral and hypnotic procedures. Extended periods of analysis that are often necessary in difficult cases allow time for further extinction and over-extinction to occur through language-mediated elicitations of the negative mediating emotions, hypnotically induced abreactions, and reciprocal inhibition through post-hypnotic suggestions. Talking is helpful only if the therapist actually stimulates clients to abandon their helpless attitude or negative expectations.

A comparison of all the psychotherapeutic schools that utilize hypnotic and/or behavioral procedures is beyond the scope of this book. However, the procedures of all successful schools of therapy are consistent with the hypnobehavioral approach. This includes abreactive extinction, generalization, spontaneous recovery, shaping, the inducement of hypersuggestibility, the implantation of positive expectations, and behavioral assignments.

When a client is decompensating because of a minor trauma, it is reasonable to look at their learning history. This is done in order to understand when and how they were sensitized to the recent CS triggering the extreme CERs mediating and reinforcing their maladaptive behavior. A client's learning history, early and overly learned avoidance behavior, and secondary gains (the gains derived from the maladaptive behavior such as attention, monetary gains, or avoidance of responsibility) are elucidatable through hypnotic procedures. The learning histories obtained through hypnotic regression support the hypnobehavioral model of A-B, AD and OCD.

Most clients were sensitized early in life, creating an anxious suggestible state and acquiring their maladaptive behavior and cognitions while again experiencing this state. Many clients also have inherited a predisposition to be anxious, and this anxiety becomes conditioned to various stimuli. Classical conditioning, operant conditioning, and the acceptance of maladaptive suggestions (attitudes and expectancies) explain the origin of the AD, OCD, and A-B. This conclusion is congruent with most therapists who rely upon hypnotic and behavioral procedures (Kroger & Fezler, 1976; Kappas, 1978). Recently, classic hypnosis procedures have been combined with meditation and cognitive behavioral therapy (Kirsch et al., 1995; Schoenberg et al., 1997; Bryant et al., 2005).

E. Behavioral Model for Obsessive-Compulsive Disorders

I have indicated that compulsive rituals can be eliminated or greatly reduced by flooding the client with CS (which elicit the urges (CERs) to commit the compulsive act) and then preventing the performance of the compulsive act. FRP was employed by Meyer (1966) in the treatment of chronic compulsives in a hospital setting. Two therapeutic ingredients, rigorous prevention of the compulsive ritual and unreinforced exposure to the overt and covert eliciting stimulus conditions, were shown to be necessary for the elimination of compulsive behavior (Levy & Meyer, 1972; Rachman & Marks, 1977; Hodgson, Rachman, & Marks, 1972). To date, FRP is the most successful method for the treatment of obsessive-compulsive behavior (Franklin & Foa, 2002, 2007)

This effective treatment is derived from Mowers' (1947) two-process theory of avoidance learning. Myers, et al., described a client who had learned that all surfaces can be contaminated; therefore the client felt contamination after touching any surface. This anxiety drove obsessive hand-washing, which only temporarily reduced the anxiety until another surface was encountered. Germs of course are everywhere, including in the air, making them impossible to avoid. While hand-washing temporarily reduced the fear, the fear quickly emerged again, locking the client into repetitive hand-washing.

The fear of contamination causes clients to isolate themselves socially, therefore restricting reinforcing experiences that brought pleasure and meaning to their lives. Also, any stressor that causes anxiety leads to these clients enacting their overly learned method of reducing anxiety by handwashing (Cromer, et al., 2007). As previously stated, avoidance responses, once learned, are extremely resistant to extinction in normal experiences because the avoidance behavior prevents extinction.

Recent evidence regarding the efficacy of cognitive-behavioral treatment of OCDs show that a majority of clients are helped by exposure techniques. Most clients experience a 50%-70% reduction in symptoms (Steketee, 1973). On a one-year follow-up, 50% report they are much improved or very much improved, and 76% have maintained their gains. 50-70% reported an improvement in the quality of their lives (Dafenback, et al., 2007).

Neuroscientists have hypothesized that OCD behavior involves a dysfunction in the serotonergic system. The exact mechanism is unknown (Lambert & Kingsly, 2005). In the 1980s I encountered a client who exercised constantly and was not responding to FRP. I sent her to Canada for Anafranil (Clomipramine) as it was not yet available in the U.S. More recently Anafranil has been shown to aid in the treatment of OCDs (Daugherty, et al., 2002). This particular client responded extremely well to the medication, and at a three-year follow-up, her improvement remained without receiving any more therapy or FRP. Generally, however, combining medical treatment with FRP does not increase the efficacy of treatment for OCD. Behavioral treatment is more effective than medical treatment alone (Franklin & Foa, 2002, 2007). Lastly, psychopharmcological treatment of OCD shows that after the medication is stopped, the relapse rates are 50% - 90% (Daugherty et al., 2007).

An excellent and detailed report of treatment for compulsive handwashing by FRP is presented by Mills (1975). He isolated baselines and included a placebo phase. If the reader wants a complete description, read Mills (1975).

Hersen and Detre (1980) report preliminary success using FRP with A-Bs. They placed a client in a situation surrounded by foods she used to binge with. She was then instructed to take a few bites of food and told to stop eating. The physiological arousal increased, as did the client's self-reports of anxiety. The patient in this environment was prevented from binge eating and inducing vomiting. FRP was continued until physiological responses returned to normal resting levels and the patient reported no further urges to binge eat. After several sessions a "temptation" test is given to assess the effects of treatment. They state, "In spite of a few successes employing this strategy, obviously further confirmation would be needed to determine whether generalization of treatment gains extend to the patient's natural environment" (p. 301).

Flooding OCs with images of feared objects and other eliciting stimuli and persuading them to engage in their most feared activities along with participant modeling was more effective than simple relaxation procedures for the treatment of OCD behavior (Rachman, et al., 1971). The research concerning FRP indicates that it is the most important variable in the effective treatment of chronic obsessive-compulsive behavior (Mills, et al., 1973; Rachman et al., 1971; Simpson & Liebowitz, 2006; Clark, 2005).

Interestingly, Mills, et al. (1973) reported that there was no consistent relationship between anxiety and the performance of compulsive rituals. Anxiety was found to be greatest before and after, rather than during the FRP procedure. Mills' observation is incongruent with what I have observed in most of the 100-plus chronic bulimics who experienced FRP during my treatment. Also, Hersen (et al., 1980) reported that urges to binge correlated with bulimics' levels of physiological arousal.

Bulimics experience a great deal of anxiety when prevented from performing their compulsive binge-eating and vomiting after eating what they consider to be a dangerous food or amount of food. As the anxiety is elicited, typical conversion reactions (e.g., feelings of getting fat in the thighs and stomach or all over, or of bloating and/or swelling) are also elicited. Those who did not immediately experience the anxiety were found to be employing the rationalization, 'I can simply wait until the therapy session is over to vomit and then get rid of these feelings - so I don't have to experience them now.

Most chronic A-Bs are skillful at dissociating negative feelings. However, even though the client may be able to postpone the anxiety, the therapist can usually elicit it by flooding the client for an extended period of time with tension-eliciting statements such as, "The longer we sit here, the more the food is being converted to fat and being deposited in your body." As the A-B faces this fact, she becomes more anxious and then usually experiences the conversion reactions. A-Bs' anxiety can be further elicited and extinguished by having her flood herself with the thoughts of getting fat as a result of having lost control and ingesting high calorie foods. The mediating emotions can also be elicited by having her imagine that she has gained so much weight that everyone is ridiculing her. This often elicits the strong feelings of anxiety (separation anxiety) which originated early in the clients' history.

Flooding OCDs with obsessive thoughts until they feel relaxed or become bored with those thoughts usually causes a significant diminution in the frequency and intensity of those thoughts. The obsessive thoughts are extinguished in the same manner that operant responses are. Since the response is elicited in the absence of reinforcement, it diminishes in frequency and intensity. As with any extinction procedure, this should be continued until the behavior, feelings, thoughts, etc. are unelicitable or are experienced as being ridiculous by the client. Again, spontaneous recovery of the extinguished obsessions is more the rule than the exception, so over-extinction is usually necessary.

Many chronic A-Bs simply defend against experiencing the negative mediating emotions by accepting the weight gain with the conscious or unconscious plan that they will simply lose the gained weight once therapy is terminated. This is often the major reason why A-Bs improve in inpatient hospital settings but return to starving or binge-eating and vomiting after returning home. This rationalization or unconscious plan has to be made conscious and be confronted.

The most effective method for making the unconscious motives conscious is through the combined use of hypnotic regression, affect bridging, separation therapy, and "trance therapy." These techniques are compatible and can be used in combination. If these techniques prove ineffectual in causing the resistant client to face her unconscious and conscious motives, direct confrontation may be helpful as some chronic resistant A-Bs are simply acting like spoiled children, angry at themselves and significant others for a variety of irrational reasons.

For example, an extremely disturbed chronic A-B who had a dissociative identity disorder was punishing her mother as well as trying to kill herself by bingeing and vomiting. Her reason was, "I'm not the prettiest girl in the world." She had been hospitalized for six months at a prestigious university hospital, but admitted to simply playing along with the therapeutic regime. The day she was released, she binged and vomited on the way home.

During hypnosis, the other personality was encountered and confronted, but no lasting positive change resulted. Her mother, who was being manipulated by this other part, was introduced to it. The mother was highly educated and not overly surprised when I introduced her to this homicidal part of her daughter's personality. She remembered encountering that personality part before, but no prior therapist had believed her. During the succeeding conversation the other personality explained to the mother exactly what she was doing. However, none of this proved very helpful, and this case proved to be a failure. On a three-year follow-up, she had not responded to any of the numerous subsequent therapies she had tried.

Another chronic A-B client, who was simply angry because she was not as pretty as she would have liked, came from a family that was wealthy, loving, and very supportive. The client vented her anger at her mother freely because she knew that her mother would not stop loving her, but felt guilty about doing this. Her bingeing and vomiting was to punish herself for her treatment of her mother. Interestingly, she would sometimes binge and not vomit in order to punish herself by making herself gain weight.

After I established a therapeutic relationship with her, she was confronted with the evidence to support the aforementioned interpretation of her resistances. She was shown it was time to "grow up" and instructed to face her attitude of self-pity and what she was doing and told there is nothing wrong with her parents showering love and affection on her, but it must be accepted with a good attitude.

She was angry when this session terminated as she did not agree with the above interpretation. However, upon returning a week later, she reported that her bingeing and vomiting had almost entirely ceased. Until this point, she had seldom been able to go three days without a bulimic episode. During this session, FRP and abreactive extinction were performed, and the client continued to progress.

As stated previously, researchers have found that many OCs do not experience increases in anxiety when their compulsive rituals are interrupted. Moreover, some even experience increased anxiety after completing their rituals. Hodgson and Rachman (1972) and Roper, Rachman, & Hodgson (1973) found that with compulsive checkers, whether or not checking reduced anxiety was dependent upon the client's "mood state" and how long the rituals took place. The longer the compulsive rituals took place, the more the client experienced anxiety. It is my opinion that longer time periods allow dissociated cognitions and emotions to surface, causing clients to experience the negative emotions which they are trying to reduce through the compulsive rituals.

Many OCs and a few A-Bs report that their anxiety is not consistently related to their compulsive behavior because they have dissociated it. Most bulimics enter trance states when performing their rituals and experience time distortion. After repeated occurrence of the trance state paired with the compulsive ritual, the combined process becomes an automatic habit. Entering trance states and dissociating the negative emotions is reinforced because the negative emotions are temporarily reduced. Following compulsive binge-eating and vomiting, the bulimic comes out of the trance and realizes what she has done. She then experiences anxiety because she realizes she lost control, which often again initiates the bingeing and vomiting cycle. The more the client has dissociated her emotions, the more out of control she becomes, causing her to experience even more guilt and anxiety.

Mood states characterized by general feelings of uneasiness increase the bulimic's tendency to automatically comply with the urge to binge and vomit. Nebulous negative emotions more often initiate overeating in obese people than do specific negative feelings (Roden, 1978). Although all types of emotional states have been observed to elicit binges, fear of the loss of control and confusion as to what they are feeling or what underlies these feelings often cause A-Bs to experience an increase in their negative mediating emotions.

In my treatment, abreactive extinction of the mediating emotions is usually employed before the FRP. After the first day of therapy, which includes abreactive extinction, separation therapy, and FRP, most clients experience significant decreases in their negative emotions and urges to binge and vomit. As proof that the therapy works, clients are placed in a situation where normally they would lose control. For example, being alone in a motel room and having to eat only a small part of a favorite food, such as a chocolate doughnut from a box of a dozen, normally elicits the overwhelming urges to binge and vomit. However, on the evening of the first day of treatment, 90% of the clients were able to overcome a situation that previously caused a loss of control.

During this exercise, the client is instructed to chart the intensity of urges and to verbally record her feelings and thoughts as she practices the FRP practices learned earlier that day. She is later instructed to do the FRP in all situations where she had previously lost control. The urges to binge and the obsessions concerning weight gain, etc., are greatly reduced (seldom lasting more than 20 minutes) following the first day of abreactive extinction with hypnosis, imagery, and FRP. FRP trials have to be carried out in the everyday environment in order to guard against spontaneous recovery and to ensure adequate generalization.

Some clients cannot or will not experience the mediating emotions under hypnosis; if this is the case, only FRP is practiced. Eliciting the negative mediating emotions solely through imagery results usually in slower extinction. However, hypnotic abreactions make the in-vivo extinction easier in addition to presenting an opportunity for the client to understand her unconscious motives, secondary gains, and how she employs dissociation to avoid emotions that need to be faced.

As clients become secure that the negative feelings, albeit extremely uncomfortable, will never physically harm them or cause them to lose their mind, they become more willing to practice the FRP procedures in the home environment. A close friend, family member, or another therapist who understands the procedures can aid the client in practicing the FRP in the home environment. This helps prevent relapses. Most bulimics binge when they are alone so they must practice the extinction on their own while orally recording their thoughts and emotions. After they become more confident, the recording can be eliminated.

Myer (1966) and Leitenberg, Rawson, and Bath (1970) summarize the important ingredients in the behavioral treatment of OCD. Myer states that:

> "Expectations of dire consequences can be countered only if both response prevention and exposure to fear stimuli are provided. Only when both ingredients are present is true extinction made possible by allowing the client to discover that the negative consequences that she anticipates are not forthcoming." (p. 158)

Leitenberg, et al., also states:

> "In many ways compulsive rituals are attempts to escape or avoid aversive stimuli. The key to rapid extinction of avoidance behavior in animals is to expose subjects to formerly conditioned fear arousing stimuli. Avoidance behavior is found to be substantially reduced following such exposure." (p. 81)

Following extinction of the maladaptive behavior, an adaptive incompatible behavior such as relaxation or another activity should be substituted for binge-eating and vomiting or other compulsive behavior. New and more adaptive coping behavior must be conditioned to occur as negative experiences are a normal part of everyday life. Post-hypnotic suggestions that initiate adaptive behavior and fortify the client's expectation of increased self-control and competence are useful in order to ensure that the desired cognitive changes persist. If a client's cognitions (expectations) do not become more positive, they may lose motivation to practice the FRP.

OCs failed to improve when they did not comply with the request to refrain from engaging in their compulsive rituals between sessions (Leitenberg, et al., 1970; Myer, et al.). Complete extinction never occurred for these clients because the maladaptive anxiety-reducing behavior was again reinforced. Throughout this book, I emphasize the importance of preventing OCs and A-Bs from engaging in any other activity that can temporarily decrease the negative emotions that mediate and reinforce their maladaptive behavior while FRP is enacted.

Similarly, with anorexics the avoidance of food cannot be allowed because the negative mediating emotions and maladaptive cognitions will not extinguish. The act of avoidance guards the negative emotions and cognitions from extinction so that mediators and reinforcers for their starvation remain. This conclusion is in agreement with Myer, et al., Leitenberg, et al., Mills, Agras, Barlow, and Mills (1973), and Bandura's (1969) conclusions regarding extinction of OC behavior.

F. In-Patient Treatment of Eating Disorders and Family Therapy

Operant conditioning in-patient behavioral approaches have shown some limited success for the treatment of A-B. However, these treatment programs usually do not emphasize the extinction of the troublesome mediating emotions that are triggered by food or stimulus situations in the home environment. Neither do they systematically attempt to modify the client's maladaptive cognitions and expectations or unconscious motives, or teach a substitute coping skill. For A-Bs, food becomes a CS that elicits a panic reaction which actively blocks consummatory behavior or elicits the urge to binge and vomit. My interpretation is congruent with Crisp's (1980) concept of anorexia as being a food phobia.

In the hospital setting, food may lose its capacity to elicit negative emotions because of the extinction that takes place as a byproduct of the operant procedures. Whenever an anorexic is coerced or shaped into eating, the act of eating will elicit the troublesome mediating emotions and cognitions, and, if she does not dissociate those emotions, extinction will occur. The operant procedure causes the client to experience the CS without those stimuli being paired with noxious situations. This causes extinction of the mediating and reinforcing thoughts and emotions that initiated and maintained starving. One would also expect the extinction to generalize to similar situations in the home environment. However, other eliciting situations present in the home environment (such as family squabbles) may be absent in the hospital and impossible to replicate; if so, these situations remain capable of eliciting the negative mediating emotions that initiate starving. The home environment is obviously different from the hospital environment on many social and physical variables. In view of the above and because few adaptive coping skills are taught in the hospital to replace the maladaptive coping behavior, recovery of the previously extinguished behavior should be expected when the client returns home.

The process of forced-feeding (hyperalimentation, etc.) also causes A-Bs to experience fear of weight gain and urges to binge and vomit which, when elicited without reinforcement, extinguish. That these procedures are anxiety eliciting is obvious from the many reports of starving A-Bs who become so anxious from the forced-feeding process that they attempt to tear out or crimp the feeding tubes. As forced-feeding continues, however, the fears are continuously elicited by the thoughts of increased caloric intake and the possibility of weight gain, and extinguish as a result of their repeated non-reinforced elicitation.

Extinction of the fear of gaining weight, loss of control, etc., in conjunction with positively reinforcing the operant response of eating are probably the important variables in the initiation of eating and weight gain in the hospital setting. However, this whole process can be blocked by the client purposely or unconsciously dissociating the mediating emotions and thoughts. A-Bs may just wear an external mask of compliance which conceals a variety of strong dissociated emotions and cognitions that, because of the dissociation, are guarded from extinction. The emotions and cognitions are forced temporarily underground, only to emerge after they leave the hospital. It is not uncommon for a chronic bulimic to report that she binged and vomited on the way home after months of insight-oriented and group therapy in a hospital setting. These observations are congruent with the findings that behavior modification programs result in a high rate of improvement in a hospital setting, but the success is not maintained when the client returns home (Bruch, 1973; Minuchin, Rosman, & Baker, 1978). Minuchin et al., concludes

> ". . . to maintain her new learning, an anorexic patient would have to generalize her responses outside the therapeutic environment. This would be possible, theoretically, in a well-differentiated context. But the anorexic is part of an enmeshed family system that encourages belonging and family loyalty and proscribes separation." (p. 89)

I agree that lack of generalization is a major reason for the relapses, but disagree with Minuchin's, et al., conclusion that "To expect the anorexic to maintain an autonomous change in the face of an unchanged family system in unrealistic" (p. 90). I have found that with chronic adult A-Bs, separation from the contextual stimuli (home), which elicits the negative behavior and thinking, can be achieved by establishing emotional distance through hypnotic and behavioral techniques. It is the strong emotions and cognitions that are elicited by the family that keep the A-B locked into the family, so it is important that these CERs and cognitions be extinguished. This is done by FRP, literal suggestions, abreactive extinction, imagery conditioning, and separation therapy.

For example, the client can be flooded with imagery of her family rejecting her until the negative emotions and thoughts are difficult or impossible to experience. The feelings of separation anxiety, fears of accepting responsibility, and expectations of failing that often are at the core of many A-Bs being unable to leave home can be extinguished by abreaction or flooding. In all cases, the many positive aspects of growing up and maintaining an independent existence are emphasized through discussions, hypnotic imagery, hypnotic suggestions, behavioral assignments, and listening to hypnotic inductions on audio recordings.

Weaning the client from the influence of the home environment is sometimes facilitated by having her move and maintain a separate residence. Often she may do this voluntarily following the extinction of her various fears and her understanding of the unconscious motives that were keeping her imprisoned in the family.

If the family system is the maintainer and elicitor of A-B behavior and thinking, then why is it maintained when the adult A-B is functioning on her own apart from family influences? Generalization of the previously learned responses to new stimulus situations is part of the answer. The covert eliciting stimuli (e.g., negative thoughts, low self-esteem, fear of social interaction, fear of growing up and leaving the family) also stay with the client irrespective of environmental changes.

Many behavior modification programs designed for the treatment of A-B completely ignore the internal conflicts and respondent aspects of the client's behavior. Thoughts, fears and conflicts are important variables and, unless modified, clients will relapse. A behavioral model that considers both the respondent and operant aspects of maladaptive behavior is reasonable (Miller & Dollard, 1950; Miller, 1975; Bandura, 1980; Eysenck, 1979; Wilson, 1982).

In summary, the association between food, gaining weight, and negative emotions and cognitions forms in a variety of ways. The end result is that food becomes a CS that elicits negative CERs and thoughts. The negative CERs elicited by food in turn block the consummatory response of eating. Again, it must also be emphasized that the conditioning process does not have to result from a traumatic event in which food was paired with a noxious situation. The acquisition of the association between food and strong negative emotions usually occurs through autosuggestion, imagery conditioning, and vicarious conditioning while the client is in a dissociative or highly suggestible state. A-Bs have accepted and rehearsed the cognition that "The reason for all my bad feeling is my weight, and if I lose weight, all my problems will be solved." This typical suggestion becomes implanted by repetition into the client''s mind and controls her behavior.

There are a variety of covert (cognitive) and overt stimuli and situations that elicit the negative mediating emotions in A-Bs. Two common situations are loneliness and boredom. During these periods suppressed feelings that mediate the typical A-B's maladaptive coping strategies emerge. People are also generally more suggestible when emotionally upset and search for quick and simple solutions to their problems. This leads the A-B to the conclusion that "Being overweight is why I am not loved and accepted; therefore I must lose weight. Because I weigh what I do means that I am worthless and unlovable. The more weight I lose, the better/more lovable I am." Self-hatred is dangerous and is characteristic of most A-Bs. Also, self-hatred is often at the core of other chronic obsessive-compulsive, self-destructive and addictive-compulsive behaviors.

A-Bs who rely heavily on dissociation are only temporarily avoiding negative mediating CERs. Therefore, dissociation is maladaptive in the sense that it prevents the client from experiencing the negative emotions and thoughts in the intensity that is necessary for them to be extinguished. As a result, the A-B imprisons herself in a vicious circle of maladaptive avoidance behavior.

The hyperactivity observed in many A-Bs serves three purposes: 1) it prevents dissociated thoughts from surfacing, 2) increased energy expenditure causes weight loss, and 3) hyperactivity blocks the experience of hunger, or at least directs the A-B's attention elsewhere and reduces anxiety.

Most A-Bs are similar to 'the true believer' as described by Hoffer (1951) in his book by the same name. A-Bs, like 'true believers,' have an extreme emotional investment in holding on to their beliefs and, when their beliefs are questioned, they simply tune out any questions.

Family members and their behavior, comments and non-verbal cues, along with correct and incorrect inferences that A-Bs derive concerning family members' attitudes, etc., may be the CS that elicit the negative mediating CERs. Also, binge eating, vomiting, and starving may be reinforced because it punished one or more of the family members, gained attention, or served as an excuse for the person to stay within the family unit and not venture out on her own. A-Bs further complicate the picture by being unaware of their motives and the reinforcers that are maintaining their maladaptive behavior. In other words, much of everyone's behavior - both adaptive and maladaptive - is habitual and so well learned that it is automatic.

Dissociation is a method by which chronic A-Bs and others sabotage therapy. This phenomenon has not been emphasized as a potent variable by behavior modifiers, but is extremely important in treatment of many chronic AD clients who are resistant to therapy. A well designed and executed behavior modification program may fail because clients guard the negative CERs from extinction by dissociating them or by unconsciously or consciously resisting extinction because of secondary gains. I have observed dissociated clients prevent extinction of mediating CERs and cognitions after as many as 30 consecutive FRPs administered over a period of 14 days. Conscious and unconscious motives and dissociation can be more powerful variables in maintaining a maladaptive behavior than behavior therapists admit. However, for clients who have a long history of anorexia, the family therapy approach, including the highly regarded Maudsley model, provides little clinical benefit (Wilson, et al., 2007).

The family therapy or systems approach apparently works well with pre-adolescent and adolescent A-Bs. However, there are disagreements as to what the significant therapeutic variables are that operate in family therapy. I contend that most of the important family therapeutic methods are explainable by less nebulous and more valid behavioral and hypnotic concepts. This does not detract from the 75% -- 90% full recovery rate after five years (LeGrange & Lock, 2005) for the treatment of non-chronic pre-adolescent and adolescent anorexics. Interestingly, the family therapy approach often employs an inpatient operant behavior modification program in order to initiate eating, with family therapy usually taking place after eating begins. Minuchin, et al., an early proponent of the systems approach, states,

> ". . . Without any doubt, when anorexia nervosa patients are treated within a year of the beginning of the illness with a systems approach in the context of their family, they can be cured in a short period of time." (p. 138)

The hypnobehavioral approach requires that the client learn appropriate methods of coping with negative environmental stimuli rather than altering the behavior of family members and others. The systems approach relies on changing other people's behavior or the system in which the client is.

This is not a reasonable solution to emotional problems for adults, however. Capable adults should be able to function in their environment or social system, and when necessary, be able to maintain their emotional distance from it. Dependence upon the alteration of other people or the environment in order for clients to live well implies that external situations and others control them. Without the clients learning new and more adaptable coping skills to control their behavior irrespective of the environment, they will remain emotionally subservient to these external forces and unable to govern their own behavior. For A-B and AD clients, their anxiety, thoughts, and behaviors are strongly influenced by too many external and internal stimuli.

On the other hand, to expect a child or adolescent to become independent of the actions and behavior of the family unit may be expecting too much. When family conflicts and behavior patterns are the mediators and maintainers of a child's or adolescent's problems, then family intervention can complement individual hypnobehavioral therapy. However, the more severely disturbed the family is, the less likely the members are to enter family therapy. This is a preselection factor, which may account for the high success rate reported by the advocates of family therapy. The systems approach and its basic philosophy that the system must change in order for the individual to change is contrary to the goals of hypnobehavioral therapy with adults, adolescents, and even children.

Family therapists must deal with the criticism that implying to clients (children, pre-adolescents, and adolescents) that the world around them should change so that they can cope is not realistic and actually dangerous. The U.S. society has spawned too many people who can do nothing but complain and expect the world to change to suit their every desire.

A 17-year-old highly intelligent boy who had been in family therapy for A-B came in from out of state for three days of therapy. He remained 100% improved on a three-year follow-up. When I first met him he had been hurt and angry at his father, a very successful businessman, because he would not spend time with his son and basically ignored him. The father was not interested in functioning as a typical father, and despite prior family therapy, was not willing to change.

However, the father told his son, "I know that I am not much of a father, but I'll pay for the best education for you that is possible." I explained to the client that he should stop begging his father to love him in the way he desired and instead accept what his father had to offer. The client understood that his father had openly admitted having problems being a traditional father, and understood the father was admitting it was because of his deficiencies. The client finally accepted that although his father was deficient in that area, he still had something to offer him.

Treatment consisted of 3 days of FRP and hypnosis to extinguish his maladaptive behavior, enhance his self-esteem, and work toward his goals. On a one-year follow-up, the client expressed that he appreciated his father being a man of his word who readily paid for his education, and was surprised by his father's compliments about his academic success. He went on to graduate from a prestigious university and was successful in a career.

Another client, a 19-year-old woman from out of state suffering with bulimia, came for three days of therapy. Her father was a well-known, powerful, and very controlling of his family. Her bulimia and cigarette smoking was her means of punishing her father for trying to control her. This client was highly intelligent and quickly realized what she was doing when it was pointed out to her. I had a conference call with her and her father, who was very loving and expressed that his control arose from being highly concerned about "the nutcases in California" who could possibly hurt his daughter. His exact statement was, "This is Charlie Manson-land." The client readily understood that her father loved her and was sincerely concerned about her, and was able to accept her father's domineering style and cope with it.

After the three days of FRP and hypnosis, she returned home free of her bulimia; on a one-year follow-up she rated herself as 100% better. This was later corroborated 10 years later by her father, who was very proud of her as she had become a well-respected medical doctor.

The major goals of hypnobehavioral therapy are for clients to learn to cope with difficult situations instead of being controlled by them, and to be reasonably free of negative CERs elicited by negative thinking and/or the environment. Real freedom is obtained when people can choose how they will react in most situations, including when they encounter family members. The locus of control should be shifted from external situations to an internalized set of adaptive coping behaviors and cognitions. This important freedom is obtained when individuals understand and apply valid behavioral principles in order to extinguish maladaptive CERs and thinking along with learning adaptive coping skills. People do not feel good about themselves when they are pawns to their environments or to other people's moods and behaviors. This new-found sense of freedom strikes a blow at what Minuchin termed the classic anorexic pattern:

> "Her refusal to eat the hot dog is a pathetic assertion of self against her conviction that she has always given in and that she will always be made to give in. This idea is a classic component of anorexia nervosa."

Short-term hypnobehavioral therapy has been approximately 50% successful for the treatment of chronic adult A-Bs who experienced repeated failure in other therapies. 70% of clients reported significant improvement. Treatment time has varied from two days to four weeks; four to six eight-hour days with periodic communication has been adequate for successful treatment with many. With those residing in proximity, the best results have been obtained after one or two days of intensive hypnotherapy followed by two to three hour sessions per week extending over a period of three months.

G. Eating Disorders: The Hypnobehavioral Model

A condition similar to anorexia in humans can be produced in mammals (Solomon, 1963). One must eat to live. However, Solomon states:

> "... Eating in dogs and cats can be permanently suppressed by a moderate shock delivered through the feet or through the food dish itself (Lichtenstein, 1950; Masserman, 1943). Such suppression effects can lead to fatal, self-starvation. A toy snake presented to a spider monkey while he is eating can result in self-starvation (Masserman and Pechtel, 1953). (p. 198)

Pairing a consummatory response with a noxious stimulus obviously has a profound suppression effect on the consummatory behavior in mammals. The physiologic correlates (visceral vasoconstriction and dry mouth) of unpleasant CERs are also incompatible with eating.

I have encountered clients in which pairing negative emotions with food had nothing to do with their relationship to the family, other people, fear of weight gain, or internal conflicts. These cases are rare but informative because they give a pure example of how pairing an aversive stimulus with fear can block a consummatory response in humans.

A client had become unable to eat only small amounts of food and drink small amounts of water because she had experienced extreme fear and anger while attempting to eat. She did not have the cognitive characteristics (e.g. fear of weight gain) or conversion reactions (body distortions and bloated feelings upon eating) that usually characterize A-B. She knew she was starving to death. Whenever she attempted to eat, she experienced a panic reaction that included dry mouth and tight feelings in the stomach, which blocked her eating. She had been in therapy with a psychiatrist for over nine months who had told her that hypnosis would not be of any help to her. Obviously her expectations for success were minimal; however, she was motivated to be rid of her problems and was obtaining little secondary gain from her behavior despite marital problems.

Through hypnotic regression she easily became aware of when and how eating and food became associated with negative emotions. Following funeral services for her mother, she became both angry and anxious with relatives during a meal. Because of her extreme state of sympathetic arousal, she gagged on the food when she attempted to eat. The gagging elicited more fear and negative emotions that simply became associated with eating and drinking. During the following year, her fears had generalized to all food and eating situations. Apparently her high state of anxiety increased her suggestibility and facilitated the acquisition of the conditioned association between eating and the intense negative emotions that mediated her choking on food and most liquids.

Through hypnotic regression and abreactive extinction, the choking, gagging, and negative feelings such as the anger and guilt that had been associated with food were abreacted and extinguished. Immediately following abreactive extinction, in-vivo extinction was conducted in order to reinforce and facilitate generalization of the extinction to real life situations. This was accomplished by having the client drink water and eat small amounts of food with me. The systematic desensitization procedure worked well, and she was instructed to continue practicing the procedure at home and in a variety of eating situations.

When this client first began therapy, she was near medical jeopardy due to a very low weight. The initial therapy involved a total of six hours; a four month follow-up indicated almost complete recovery. Approximately one year later she experienced a minor relapse; this was reversed after two additional hours of repeating the same procedures.

It must be emphasized that this is not a typical case of anorexia nervosa. She did not voluntarily begin restricting her food intake and had no unconscious or conscious motive for starving. She was neither afraid to gain weight nor did she experience a distorted body image. This case clearly illustrates how negative emotions can be paired with a consummatory behavior and block that behavior. I have successfully treated at least six other similar cases which were originally mistakenly diagnosed as anorexia nervosa and, similarly, treatment time was very short.

An emotional state can serve as a punisher for an operant consummatory response. Humans possess covert processes that serve as CS that can elicit strong CERs. These CERs can serve as punishers for a consummatory response such as eating (Kroger & Fezler, 1976).

While working at the Midtown Medical Center in Atlanta, Georgia, I accidentally produced anorexia nervosa in an overweight, hypersuggestible client. The client was instructed while hypnotized to use her imagery-conditioning skills to associate negative emotions with junk food. The goal was to have junk food elicit negative feelings such as mild guilt and fear. The end result hoped for was that these negative emotions would block the consumption of fattening foods.

She began to lose weight steadily and continued the program on her own. However she proceeded to generalize her negative feelings to a variety of foods by using her self-hypnosis skills to pair feelings of low self-esteem and guilt with gaining weight, and feelings of high self-esteem and a positive self-image with starving. Therefore, whenever she began to feel insecure she stopped eating, which reduced her negative feelings. She also began to manifest the typical A-B conversion reactions whenever she ingested food. These reactions included sensations of bloating, and were exacerbated by auto-hypnotic and hetero-hypnotic suggestions that she would feel full and more satisfied on less and less food.

When she returned after a number of months, she was pleased with her extreme weight loss and, like most A-Bs, did not view herself as overly thin even though she had become emaciated. The process was easily reversed by using hypnosis to have her imagine a healthy body image along with normal eating habits. Her weight steadily returned and ultimately stabilized at a normal healthy weight. She remained stable on a four-year follow-up.

A-Bs often do not lose their hunger until later stages of starvation. In fact, they are often continuously hungry. Following periods of starvation most A-Bs crave food but are unable to eat because of the CERs and conversion reactions that are elicited by food. Of course, the conflict between the feelings of hunger and the fears of eating generate high levels of anxiety that goad A-Bs into more binge-eating and vomiting and/or more starvation and dissociation. As starvation increases, sensations of hunger may disappear entirely as a result of vitamin deficiencies. Roger Williams (1976) stated, "It is well established that withholding a single essential vitamin (vitamin B1) from fowls, rats or human beings will cripple the working of their appetite mechanisms" (p. 126).

When A-Bs reach this state, medical intervention involving forced feeding may become necessary. At this point an emotional problem has clearly caused a medical problem. During my treatment, all clients are strongly urged to take at least a multiple vitamin. Many have shown serious nutritional imbalances, I always urge A-Bs to obtain a nutritional analysis by a physician who specializes in nutritional therapy. Most clients comply with the vitamin protocol when they realize that malnutrition can cause electrolytic imbalances, bloating, and water retention, which make them feel and look heavier.

H. Cognitive-Behavioral Treatment of Depression

Chronic A-Bs become so hindered by their fears of rejection, low self-esteem, compulsions, and fear of food that they avoid activities that would normally be enjoyable. When reinforcers are withdrawn, avoided, or lose their effectiveness, the frequency of the emitted or voluntary behavior that they reinforce diminishes. Many of these behaviors are those that added meaning and enjoyment to their lives. Lack of participation in these previously enjoyable activities results in them becoming dyseuphoric.

Biochemical changes in the brain take place as a result of the attitude of helplessness and stress, and maintain and deepen the depression (Weiss, et al., 1979; Schildkraut, 1969; Levor et al., 1980; Affleck et al., 1994; Keefe et al, 2002). Chronic stress can damage brain cells especially in the hippocampus (Sapolsky, 2000). The physiologically damaging effects of chronic stress and its contribution to a variety of mental disorders is beyond the scope of this book. However, one major conclusion is that all psychotherapeutic treatment programs should employ a stress reduction-relaxation program such as Jacobson's progressive relaxation, biofeedback training, and hypnosis relaxation training. Teaching clients how to relax quickly and reduce stress takes away some of the negative power of the CERs.

It has been my experience that depressed clients respond well to vasodilation biofeedback training in conjunction with the behavioral procedures outlined later in this chapter. Increasing parasympathetic activity through biofeedback vasodilation training probably aids in the restoration of biochemical imbalances that occurred as a result of stress, by reciprocally inhibiting the sympathetic response.

Cognitive models of depression have been presented by Beck (1967, 1974) and Seligman (1968,

1973, 1974, 1975). Maier and Seligman (1976) extended the learned helplessness model initially derived from research with lower animals to explain human helplessness and depression. Animals who have learned that their behavior is independent of obtaining reward or escaping punishment show behavior similar to human depressives. Clients who feel helpless, i.e., who believe that their behavior does not increase their probability of being rewarded, manifest passiveness and unwillingness to initiate responses, just as do helpless animals. Also, people who feel helpless expect negative consequences when they do act, which often prevents them from responding.

Non-contingent positive reinforcement (rewards that do not systematically follow behavior) can explain the origins of some of the indifferent, pessimistic and helpless attitudes encountered in clients who were reared in affluent homes. The good things in life such as a car, vacations, and praise, are often given without the young person doing anything purposeful or constructive. Again, the behavior we see is usually what has been reinforced. As a result, many of these young people develop the maladaptive expectation that they deserve the good things in life without working for them and, as a consequence, feel cheated when they do have to work for them. When these individuals enter the adult world, their negative attitudes cause the probability of being rewarded to decrease, leading to a deeper sense of helplessness. The end result is that these people become passive complainers who feel they are entitled to the good things in life without working for them. Non-contingent reinforcement, e.g., giving money and rewards when no adaptive behavior has occurred, can cause people to acquire an attitude of apathy and a sense of entitlement. These spoiled people are often the most difficult to work with because they avoid all suffering at all costs. My approach requires some suffering.

Non-contingent positive reinforcement can also explain the helpless attitudes encountered in many welfare recipients. They are given a minimal amount of reinforcement (money and medical care benefits) that barely keeps them alive. When they do try to work, benefits that are necessary for their children or themselves are cut or discontinued. Since benefits like medical care are often not provided for by low-paying jobs, they often find they are better off not working. This puts individuals into a situation that could not be worse. Positive reinforcement is received for not working and negative reinforcement (taking reinforcement away) is received for working.

Seligman (1974) maintains that it is not traumas that produce the helpless behavior, but instead it is the lack of control over traumatic situations. Depressed individuals feel that their behavior is ineffectual. Seligman believes that the internalization of the cognition that 'What I do doesn't make a difference' is important for the origin and maintenance of depression. Since 1982 Seligman's learned helplessness model has generated interesting research, proving core validity. In the 1960s Seligman and colleagues (Maier, Seligman, & Solomon, 1969; Overmier & Seligman, 1967) exposed dogs to uncontrollable shocks. The dogs, not being able to make an avoidance response, became passive and helpless and did not emit avoidance behaviors that could be reinforced. Other dogs exposed to equal amounts of shocks in situations where they could control the shocks readily learned avoidance responses. The conclusion was that when animals learn that they have no control over aversive events they enter a state of helplessness, which causes them to be unable to exhibit adaptive coping behavior and be untrainable. Seligman also demonstrated learned helplessness in humans (Hirota & Seligman, 1975).

Abramson, Seligman and Teasdale (1978) reformulated the theory to better explain human depression by considering the cognitive components of helplessness. An important cognitive variable is how people who are suffering aversive events explain the reasons for those events. Abramson, et al., concluded that people who have internalized a consistently pessimistic attitude are vulnerable to depression when faced with uncontrollable negative events. The hopelessness theory of depression

(Abramson, et al., 1989) proposes that a hopeless expectation (the belief that one has no control over what is going to happen, bad outcomes will occur, and highly desired outcomes are not going to occur) is essential in maintaining depression.

Depressed people are not relaxed, but chronically anxious. The autonomic nervous system correlates of sympathetic arousal are present in most depressives. Depressed people are not relaxed but experiencing negative arousal. For example, their hand temperatures are chronically low, indicating sympathetic activity. Again, I must emphasize the importance of relaxation training to reduce this anxious hyperarousal. That anxiety and depression co-vary has been supported by self-reports, clinical ratings, diagnoses, and genetic influences (Watson, 2005). Barlow (2002) and Mineka (et al., 1998) have recognized that anxiety is a major risk factor in both unipolar and bipolar depression. Therefore, since many chronic A-Bs experience significant negative affect, treatment which combines stress reduction with treatment for depression should be an integral part of any treatment.

Most A-Bs, ADs, and depressives have expectations that their efforts will not improve their behavior and believe that they are victims of their own uncontrollable urges, behaviors and fears. This conviction deepens their dyseuphoria, strengthens their sense of helplessness, and explains why many chronic A-Bs and ADs are depressed.

Beck (1964, 1974) proposes that an individual's negative perception and appraisal of environmental events sustains depressive behavior. Beck's theory, like Seligman's, hypothesizes that cognitive variables are of paramount importance in the development and maintenance of depression. In Beck's theory, negative expectations concerning the future and a negative perception of the self and the world are central to the depressive's cognitive set. The cognitive distortions maintaining depression are: 1) the use of arbitrary inference (reaching conclusions while ignoring relevant evidence), 2) selective abstraction (fixating on one detail of a situation while ignoring the context), 3) over-generalization, 4) minimization of evidence contrary to their negative set while maximizing supporting data, and 5) perceiving impersonal events as personal affronts. These cognitive distortions must be confronted and altered.

The concept of locus of control is important in understanding A-B and AD. Mandler (1966) suggests that the feeling of being out of control is a central characteristic of all ADs. Seligman, et al., suggests that anxiety is the initial response to a stressful situation, and depression begins when a person internalizes the belief that he is helpless or out of control and the control is unattainable. Beck's (1967) cognitive theory of depression is similar to Seligman's theory in that he also emphasizes a client's perception of control as an important variable. Along with Bandura (1980), all of the aforementioned authors apparently agree that cognitive variables concerning how competent or powerful a person feels has a great influence on the development and maintenance of "healthy" behavior.

The question of control is also a central concern for most A-Bs. The following quote from an A-B illustrates this point: "My binge-eating and vomiting are my weapons and the only way I can control things -- and you want to take them away." The fear of loss of control and the attitude of helplessness, etc., in conjunction with an impoverished lifestyle and strong self-destructive motives maintained by secondary gains, can lead to a resistant depressive behavioral pattern. The fear of losing control and a negative self-concept often intensifies the negative emotions that clients avoid or temporarily suppress through maladaptive behavior. The increase in the negative mediating emotions causes an increase in the urges to commit the maladaptive behavior. The stronger the fear of loss of control, the greater the probability of losing control. A vicious positive feedback loop is now formed.

Fear of loss of control and urges to binge-eat or vomit can usually be elicited and extinguished by having A-Bs imagine while under hypnosis they are in situations where they usually lose control. As the fear of loss of control and the urges to binge or vomit intensify, the typical conversion reactions of bloating and feelings of immediate weight gain and/or swelling or bloating in the face, stomach, hips, and thighs usually emerge. The intensity of the conversion reactions parallel the intensity of the fear of losing control and other mediating emotions and urges.

As stated, many chronic A-B and AD clients experience impoverished lifestyles. Their negative emotions keep them from venturing into social and work situations where they could learn the necessary skills to have a meaningful adult life. Clients who have restricted their lives to staying at home and being taken care of by parents, or who have few social or work skills and are not motivated to improve, are poor candidates for my short-term treatment. The secondary gains derived from their maladaptive behavior are often more powerful than the motivation to improve, so the maladaptive behavior is held on to tenaciously.

For example, I encountered a client in her early 30s who had stayed at home for over 13 years without ever working. Her parents felt trapped by her condition and supported her while she lived on the brink of starvation. As a result, she developed few social skills and no work skills. She did realize that in order to lead an independent adult lifestyle, she would have to work at jobs that she felt were below her social position. She also realized that to achieve the professional skills that were needed in order to maintain the lifestyle she wanted, she would have to work and strive like everyone else.

However, she stated she was unwilling to do this as she knew her parents were wealthy, and she felt it would involve too much work and sacrifice on her part to function as an independent adult. This gave her little motivation to improve. During therapy, her fears of weight gain, etc., were repeatedly extinguished through abreactive extinction and FRP, and she reported significant diminution of her social fears and urges to binge during and immediately after therapy. Despite this, initially there was little lasting change in her eating behavior; the secondary gains of avoiding adult responsibilities at home were reinforcing her maladaptive behavior.

Her parents were helped to see through her emotional blackmail and suicidal threats and to understand the secondary gains. Once these were understood, they refused to be further manipulated. Following this meeting with her parents and after seven days of therapy spread out over three weeks, the client was forced into living separately from her family. Her parents held her accountable despite her suicide threats and repeated attempts to return home. She eventually became partially self-supporting and reported her life is more enjoyable. However, she infrequently still engages in bulimic behavior.

A long-term behavior modification program that reinforces and shapes appropriate responsible behavior is the only reasonable treatment for this type of client. I do not define long term as 3 years of trance therapy (Schwartz, 1994). These clients need to formulate short-term goals (3-6 months), long-term goals (1-2 years), and have a life coach to keep them on track. How much money one needs to earn to have a worthwhile life must be emphasized. I believe there's nothing wrong with parents economically helping an adult child who is working toward worthy goals. although monetary gifts should of course be contingent on the recipient maintaining a healthy lifestyle, working hard and being productive.

Some clients feel proud of their adult behavior after being forced into becoming self-supporting, which in turn mediates and reinforces more responsible behavior. When clients realize their A-B behavior is no longer getting them what they want, they are willing to give up their maladaptive behavior.

Most chronic A-B and AD clients have a deficient repertoire of coping skills. Only a minimal number of social and work skills may be necessary for clients to realize that with effort and work they can derive meaning and enjoyment from their lives. The social, work, and adaptive coping skills that clients have previously acquired should be acknowledged and built upon. When people derive benefits from coping in an adaptive manner, they will more easily internalize the expectation that, "What I do *can* make a difference in what I get out of life." Once clients develop this attitude, they are more willing to learn new coping skills and become motivated to practice generalization of adaptive coping behavior and FRP on their own to eliminate the mediating emotions and maladaptive behaviors that have prevented them from experiencing a more meaningful existence. An adaptive positive feedback loop is thus formed: more adaptive coping behavior produces more reinforcement, which in turn produces more adult coping behavior.

As previously stated, most chronic A-Bs and ADs exhibit significant depressive cognitions and behaviors. In view of the fact that 15% of all adult Americans manifest significant depression and that it is probably the number one mental disorder (Secunda, 1973), clinicians should consider its treatment as an integral part of the treatment plan for most chronic AD and A-B clients.

The following is a short review of other behavioral and cognitive theories and treatment methods for depression that I have found useful. The various methods deduced from the theories are compatible with the hypnobehavioral approach. This review is not intended to be comprehensive, but instead to illustrate the theoretical and methodological basis of my treatment for depression.

Most researchers accept the possibility that there may be biologically based depression and that heredity and some chronic infections may play a significant role in its origin. The biochemical differences between normals and depressives may, however, be the result of the depressive behavior pattern and not just a result of heredity. I have found that an improved diet which takes hypoglycemia into consideration is a significant therapeutic variable in the treatment of depressed clients.

Although hypoglycemia may be effectively handled by a psychologist and a consulting physician, the role of a competent orthomolecular physician can be indispensable concerning other nutritional variables. Nutritional variables are reviewed in Chapter Q: *Non-Psychological Variables*. More extensive reviews are also recommended.

Ferster (1973, 1974) and Lewinsohn (1974a, 1974b) have focused on a functional analysis of depressive behavior and view depression as a learned maladaptive behavior characterized by deficient social skills as a common precursor. The functional approach assumes that a thorough description of the maladaptive behavior is necessary (Kanfer & Suslow, 1965). Altering the maladaptive behavior involves the manipulation of the relevant controlling variables that were identified and validated through the functional analysis. Empirical observations of depressives show a correlation between certain activities that clients engage in and the specific emotions they experience.

My observations support Lewinsohn's and Ferster's conclusions. I have seldom encountered clients complaining about depression who were actively engaged in activities that they enjoyed and found meaningful and/or were progressing toward goals. People who are self-actualizing and interested in a variety of activities seldom manifest depression. Enjoyable, interesting activities add meaning to one's life. Depressives have either stopped pursuing or have not yet acquired these activities. When people have only a few activities that are enjoyable and they are taken from them, depression often results.

I must emphasize that many MHPs and medical professionals assume immediately that someone who is fatigued and has stopped doing rewarding activities is depressed. Often people are depressed because they have a chronic physical illness that may or may not involve pain. Chronic fatigue is part of most illnesses and is part of the body's attempt to protect itself. These people become depressed because they are continually being robbed of the energy needed to take part in their former lifestyle. My policy was to continue to encourage these people to look for a medical diagnosis. The history of medicine is riddled with many horror stories of MDs and MHPs attributing what were clearly disease symptoms to mental states and psychologizing them to death -- at times literally.

In one case, I was referred a young woman diagnosed as bulimic. Her MD and psychologist stated she was a clear case of bulimia because she was slim, attractive, and vomiting repeatedly for no obvious reason. The initial interview indicated that she had none of the cognitive and emotional characteristics of a bulimic. She insisted that she felt ill and desired to regain weight as she was too thin. After many problems with her internist and former psychologist, I continued to treat her and also send her for medical workups. I did no FRP, etc., with her; only hypnotic relaxation recordings were made to help decrease the stress caused by these other health professionals and her fears of being seriously ill. Within two months of seeing her, a competent internist found she had brucellosis. This explained all of her symptoms-- and at a year follow-up, she was at a normal weight and of course no longer depressed.

Another woman was referred to me for what another internist thought was conversion pain in her foot. This client was also depressed, as anyone would be when suffering pain. During the initial interview, the client explained that the pain followed a car accident, and that her foot was swollen. The internist knew about the accident but kept ignoring the client's complaints and instead believed the pain and swelling were emotionally based. I encouraged her to consult another physician. After an X-ray, which had never been performed by the internist, it was found her foot was broken. At the time, I was working alongside a neurologist who warned me that I had made an enemy of the original referring doctor because I had referred the patient to a different doctor. I find this type of "medical fraternity" appalling as clients' suffering is not taken into account, only the saving of face for the doctor(s) involved.

Negative emotions and maladaptive attitudes and beliefs often mediate clients' withdrawals from pursuing enjoyable activities. Perfectionistic-compulsive clients may fear rejection and/or failure, and avoid all challenging situations. They may also express self-hatred because they have not lived up to societal or parental expectations, resulting in further self-punishment. Frequent aversive events and the loss of rewarding activities may all further contribute to the onset of depression. The central emphasis in this interpretation of depression is socio-environmental (Becker, 1977).

In summary, the main difference between depressives and non-depressive is that depressives exhibit a deficiency of instrumental skills needed to obtain reinforcement. As reinforcer effectiveness decreases or reinforcing activities are lost, inactivity and an internalization of the attitude of helplessness ensues. The functional approach to understanding of depression has yielded important correlations. Depressed people emit fewer behaviors (Shaffer & Lewinsohn, 1971; Libet & Lewinsohn, 1973), obtain less reinforcement (MacPhillamy & Lewinsohn, 1971), and are more sensitive to aversive stimuli (Lewinsohn, Lobitz & Wilson, 1973) than normals. Depressed individuals also have fewer social skills (Libet, et al., 1973; Ingram, Scott & Siegle, 1999) and exhibit deficits in non-verbal communication skills as compared to normals and psychiatric controls.

The treatment method derived from a functional analysis involves assessing what conditions and activities parallel or precede dyseuphoria and negative thinking. Once target behaviors and their antecedents are defined, the treatment consists of: 1) eliminating the controlling variables, and 2) initiating and reinforcing adaptive behavior that is incompatible with depression.

Clients must keep a log of their activities and chart where and when they experience depressive thoughts and feelings. The log also must include what and when clients eat, take medication, or binge and vomit. Log-keeping is important because clients' memories of the conditions present when they were experiencing negative affect are inaccurate. The client and therapist review the log and devise a strategy to eliminate the negative behaviors and eliciting situations, and to substitute adaptable and incompatible behavior for the unadaptable behavior. The log also aids the therapist in identifying relationships between the ingestion of food and other substances and mood. Hypoglycemia and/or occasionally food allergies may trigger physiological reactions that exacerbate depression and anxiety, and this can be seen by reviewing the log. The use of a behavioral log is compatible with the other theories of depression discussed below. It is always helpful to obtain accurate information as to what the client is doing rather than relying on often distorted perceptions and verbal reports.

Stampfl and Levis (1969, 1976) combined Mowrer's (1947, 1960) two-process theory of avoidance learning and Miller's (1959) conflict theory in order to explain the origin of depression. Depression is viewed as originating in two ways.

The first proposes that depression results in a complex combination of a loss of positive emotion with anxiety arousal. The basic idea is that a depressed person has learned a depressive behavior and been reinforced for it. Negative CERs act as mediators and reinforcers for depressive behavior. This model is, of course, similar to the behavioral model for A-B and other behavioral models for depression.

The second is based on a conflict multi-process approach-avoidance model. This view suggests that depressive behavior begins when anxiety has been conditioned to cues which precede punishment for participation in a forbidden act. Repeated punishment for committing the forbidden act simply increases the level of anxiety associated with that act. However, when the person commits the act, especially one with primary reinforcing values such as sex or eating good-tasting food, the punishment blocks the completion of the behavior. This leads to frustration and anger. As anger is usually punished, the person inhibits her anger. The depressive reaction serves the purpose of preventing a full exposure to the aversive situations and, therefore, extinction of the emotions is prevented. Again, it appears that a major class of mood disorders are avoidance behaviors.

The aforementioned theories are compatible and consistent with my clinical observations. However, they have different emphases and advocate somewhat different treatment approaches. The major division between the theories would be to label Seligman's and Beck's theories as 'cognitive theories,' whereas Lewinsohn's and Stampfl's theories would be labeled as 'behavioral theories.' The cognitive approach emphasizes that behavioral changes will occur as cognitions change, whereas the behavioral approach emphasizes that cognitive change usually follows behavioral change. Treatment strategies derived from both approaches have been shown to be compatible and helpful in the treatment of depression and should be combined.

Cognitive behavioral therapy is the combination of the above. It teaches clients to examine and alter automatic or overly learned negative thoughts such as unnecessary guilt, feelings of worthlessness, and self-loathing. Challenging negative thoughts along with thought-stopping techniques helps clients challenge these cognitions that maintain depression. Post-hypnotic suggestions, described in Chapter O: Laws of Successful Living, are very helpful.

I. State Dependent Learning and Generalization

The phenomena of state-dependent learning (Girden & Culler, 1937) demonstrates that responses are learned to both external and internal stimuli such as the organism's physiology and emotional state (Overton, 1969). Therefore, if these stimuli are not present at recall, learning may not be demonstrated.

Girden, et al., gave dogs a curare-type drug which paralyzed the striate musculature. They found that conditioned autonomic responses learned while the dog was drugged disappeared in the non-drugged state, but reappeared when the dog was again drugged. The responses were learned to a set of internal stimuli which were produced by the drug as well as to the external stimulus situations. Girden found that a dog could be trained to make one response to an external stimulus while in the drugged state and a different response to the same stimulus while in the non-drugged condition.

The concepts of state-dependent learning and failure to generalize are for practical purposes the same. Once a response has been taught to occur to a particular CS, a similar stimulus will also elicit that response, and the magnitude of the response will vary as a function of the similarity between the original CS and the novel stimulus. This holds true for both internal and external stimuli.

If an organism in a drug state learns a particular CR and the drug state is altered, the magnitude of the response will diminish as a function of the dissimilarity between the original drug state and the novel state. Generalization holds true for both internal and external stimuli. In other words, if an organism in a drugged state learns a CR and the drugged state is subsequently altered, the magnitude of the response will diminish as a function of the dissimilarity between the two physiological states. Many examples of the influence of state-dependent learning on the retrieval of a memory can be cited. For example, an alcoholic who hid a bottle when intoxicated likely will not remember where it was hidden when sober, but upon re-intoxication he may easily remember where the bottle was hidden.

State-dependent learning leads one to expect that the therapeutic effect of abreactions elicited under the influence of drugs would not generalize well to the non-drugged condition. Thus, abreactions elicited under the influence of drugs may yield little therapeutic benefit (Hordern, 1952). The extinction process involves learning and is specific to the stimuli present when it takes place. Therefore, when extinction is performed while an organism is in a drugged state, generalization decrements are again expected when the drugged state is gone. As previously stated, the generalization decrement is related to the dissimilarity between the drugged state during which the learning occurred and the drug-free state.

Abreactive extinction under the influence of a drug is usually carried out in a doctor's office, resulting in extinction being associated with the context in which it occurred as well as the drug state. If there is no drug state, considerable generalization of extinction would be expected to the normal environment and especially to specific internal stimuli, thoughts, and so forth, which clients carry with them everywhere. Therefore, significant generalization should occur even if these responses were extinguished only in the therapist's office.

For example, anger is much the same in the therapist's office as it is in other environments. Having the client imagine anger and its extinction in the office will generalize well to a variety of everyday situations. However, the most lasting results are obtained when clients understand the extinction procedures and practice them in-vivo in as many different environments as possible. Over-extinction should also be practiced to insure against spontaneous recovery.

In summary, responses (overt behavior and cognitions) are learned to the stimulus environment (contextual associations) and to the stimulus characteristics of a particular state (state-dependent learning). Whenever extinction occurs or coping responses are taught while a client is under the influence of psychotropic medication, a diminution of this newly learned behavior is expected when the client stops taking the drug. When a client is being withdrawn from the psychotropic medication, the conditioning or extinction procedures should be continued to counteract the state-dependent learning and the spontaneous recovery of previously extinguished behavior. If clients are not warned of the state-dependent learning effect and spontaneous recovery, they may lose faith in the therapeutic regime, question the competence of the therapist, and form negative expectations that may impede therapeutic gains when the previously extinguished behavior returns or newly learned coping behavior fails in their everyday environment.

A thorough explanation of therapeutic procedures, state-dependent learning and generalization helps establish rapport. Most clients appreciate when clinicians take the time to explain the basic ideas underlying the treatment. All too often explanations of therapy are given in terms of vague generalities by MHPs, which may only confuse clients or cause them to question whether or not therapists know what they are doing. Explanations that predict what will happen and make sense out of chaos help increase compliance and suggestibility and build positive expectations by increasing the clients' faith in the competency of therapists and the therapeutic procedures.

At this point the reader should read Case Study I as it illustrates how behavioral interventions were used to extinguish a variety of CERs and avoidance behaviors. My routine use of hypnosis on the first day was apparently useless-- or going to take too long-- so I abandoned it and instead began behavioral procedures.

J. Anxiety, Dissociation, and Conversion Features of Anorexia Nervosa and Bulimia

Understanding the origin of the emotional states that mediate maladaptive behavior is of theoretical, practical, and heuristic importance. Most people experience unpleasant emotions such as anxiety, guilt, or anger that may mediate and reinforce maladaptive behavior. However, it is to the degree which our social and emotional functioning is impaired by these emotions and the maladaptive behavior that determines whether or not we are labeled emotionally disturbed or well-adjusted.

Anxiety can occur as a result of a variety of situations that are physically or emotionally traumatic such as injuries, rejection from significant others, or early separation from the security of the home. These anxieties generalize to similar stimulus situations encountered throughout the person's life. The physiological correlates of anxiety (over-sympathetic activity and general somatic muscle tension) are similar for all people and also follow the generalization rule.

During an emotional upset, one will experience vasoconstriction through the arterial tree, increased blood pressure and heart rate, cold clammy extremities, cessation of normal digestion and pupil dilation. Anxiety also causes tension in particular muscle groups throughout the striated musculature, especially the upper trapezius and masseter muscles. Generalization of these responses can explain why clients report feelings of an increased heartbeat and other physical signs of tension in environments which previously had not directly been paired with anxiety.

Specific response systems may be more reactive in one person than in another. While an individual's physiological response profile to anxiety may differ from another's, it will be highly consistent for that individual throughout her lifetime (Lacy, 1963). For example, if a client responds to anxiety by experiencing a gastrointestinal upset, then she will respond to stress in a similar manner throughout her life. To what degree an individual's response profile is determined by learning or heredity is open to speculation.

Children are necessarily suggestible and rapidly conditionable, so it is reasonable to assume that most emotional problems are acquired early in life. Children are often sensitized by traumatic experiences early in life, and later traumas trigger or add to their latent anxiety, causing the anxiety-mediated maladaptive behaviors to emerge.

Breger's (1974) account of the origins of anxiety is consistent with my clinical findings concerning the origin of anxiety in most A-Bs, OCDs, and ADs. He derives support for this theory from numerous ethological as well as psychological studies, and concludes that separation anxiety is a prototype for all later anxiety (Bowlby, 1980).

Breger states:

"... anxiety appears very early when the attached infant becomes separated from his mother. Separation anxiety is an important motive in maintaining the mother-infant closeness, so important for the infant's survival. It is the primary source of interpersonal anxiety. The basic prototype of anxiety is abandonment." (p. 89)

The mediating anxiety that A-Bs and OCDs experience is usually identical to the same strong negative emotions that children feel when separated from their parents. Through hypnotic regression, most clients can be regressed to these emotions and to the various ages when the feelings of abandonment, loss of love, rejection, etc., were encountered. Most clients report that the negative feelings that they currently experience are identical to those encountered during hypnotic regression. This supports Breger's (1974) theory concerning the origin of neurotic anxiety. I feel that Breger's theory is the best explanation for the origins of the negative emotions which underly the maladaptive behavior exhibited by most A-B and AD clients, as well as many phobics who have not experienced other specific traumas.

'Existential anxiety' (a sense of meaninglessness or loss of sense of reality and depersonalization), which is discussed by existential writers, is the result of a person dissociating feelings, behavior, and thoughts to the extent that the security of his basic identity is disrupted (Breger, et al.). Existential anxiety is the same as the anxiety that most psychological theorists discuss. Clients face the anxiety, dissociate it, and cope with it by responding in a variety of maladaptive and adaptive ways. In order to explain the anxiety, clients (and many therapists) try to analyze it, rationalize it, and put a variety of labels on it. However, explaining and understanding anxiety does not eliminate it. The extinction process is the most understood and reliable way of eliminating the association between this basic emotional state and the stimuli, which have come to elicit it through conditioning. The stimulus response associations may never be broken, but may be permanently inhibited through active inhibition.

It has been my experience that existential anxiety can be eliminated through the application of a combination of Seligman's and Lewinsohn's procedures for the treatment of depression. As previously stated, vasodilation (thermal) biofeedback training, or any method which teaches clients how to use their imagination to increase parasympathetic activity, greatly reduces clients' baseline anxiety and accelerates the generalization of relaxation and a sense of self-control. When available, these procedures should be used with all anxious clients as their anxiety has become a strongly conditioned and generalized habit. Also, progressive relaxation training is helpful as are hypnosis audio recordings. There are apparently no contraindications to these procedures and, therefore, they should be used routinely. The interested reader is referred to the extensive theoretical and clinical literature concerning thermal biofeedback training and relaxation training.

Feelings of helplessness and clients' lack of participation in reinforcing activities are at the core of the existential crisis. The extinction of those feelings and the ability to relax strikes a deathblow to this core. The initiation and shaping of clients' participation in meaningful activities decreases the verbal reports of existential anxiety and sense of meaninglessness. Obviously people who find their lives meaningful are involved in what they consider to be meaningful activities.

I have never treated anyone for existential anxiety or depression who believed that they were competent, had goals that they were working toward, or believed that they were involved in interesting activities. Active engagement in interesting and purposeful activities is what adds meaning to one's life. Whenever therapists encounter depression, they should prod clients into activities previously meaningful and rewarding, or those which they believed may be interesting. Venturing into new worlds of activity helps clients gain a sense of mastery which in turn facilitates the extinction of anxieties. In most cases, these anxieties have been causing them to avoid those pleasurable activities. As clients again become active and regain lost skills or acquire new ones, life becomes more enjoyable and this reinforces more of the same efforts. Also, they may begin to adopt a more active and assertive attitude, which further increases their self-esteem. An adaptive self-reinforcing feedback loop is now formed, maintaining and solidifying their participation in a rewarding lifestyle.

After a few months of this type of treatment, most clients remark that they have not thought much about their existential problems and that they now view their lives as having meaning. This "sense of meaning" is a by-product of a particular type of lifestyle and attitudinal set and not the result of analysis and insight into how and why a client acquired the maladaptive thinking.

The above supports the behaviorist position that changing the behavior of a person causes a change in cognitions, whereas changing the cognitions seldom results in lasting behavioral changes. From an adaptive perspective, it is reasonable that many behavioral mechanisms are dominant over cognitive mechanisms. Rewarded actions or behaviors are more adaptable than correct thinking not followed by adaptive behavior. It therefore seems reasonable that adaptive, i.e., rewarded, changes in the behavioral system would be a potent force in altering the cognitive system, whereas the reverse may not be true.

The fact that hypnotic suggestion has been repeatedly shown to mediate change in behavior is incongruent with the position that cognitive change seldom causes behavior change. Hypnotic suggestion obviously works through the cognitive-verbal system and, because hypnotic suggestions result in behavioral changes, this contradicts the absolute acceptance of the aforementioned behavioral doctrine (Barnier & McConkey, 1998a, 1998b). It has been my experience that unless the behavioral changes mediated through hypnotic suggestions are reinforced, the old habitual ways of coping will return. This does not mean that direct hypnotic suggestion is not an important tool; on the contrary, it is often the most efficient method of mediating behavioral change, and if the behavioral change is reinforced by its consequences, it remains.

In summary, intellectualizing and talking about various philosophical problems with clients seldom mediates adaptive behavior. The verbal response system often functions independently of the more primitive affective system. Cognition and verbal behavior are separate response systems from emotional and operant behavior, and changing one (especially the verbal-cognitive system) may not result in a change in the respondent or operant system (Mikulas, 1974).

The effective treatment for existential anxiety is the same as for depression. The negative mediating emotions that cause the maladaptive behavior and reinforce it must be extinguished, and new coping skills must be taught. Increasing clients' sense of mastery over themselves and their environment by having them learn coping styles involving challenging life's problems adds meaning to their existence and increases their sense of self-esteem. Realistically, high self-esteem, a sense of mastery, and low anxiety are all incompatible with AD and the existential crisis.

An adaptive personality is one in which few maladaptive coping strategies are employed and the person is secure and willing to confront life's challenges. Well-adjusted people have reasonably high self-esteem, little anxiety, feel in control most of the time, and are able to effectively govern their own lives. The majority of their behaviors are congruent with their verbalizations, emotions, and desires. They are consciously aware of their feelings and motives, and have a realistic conception of how their culture and the real world operate. The world is perceived as neither malevolent or kind, but as usually fair along with the realization that bad luck can occur. Other people's actions toward them rarely have the power to send them into an emotional tailspin because they value and respect themselves and feel reasonably secure that their opinions are valid. People who are well-adjusted are able to form loving relationships and are tolerant of the actions of others as long as those actions do not jeopardize their or their loved ones' physical or emotional well-being.

To put it more simply, well-adjusted people cope with their problems and the outside world in an active, confident, and adaptive way. They possess a set of adaptive coping skills that they readily employ when difficult situations are encountered. Successfully mastering various difficult situations reinforces use of these skills and, as time goes by, well-adjusted people resort to these adaptive coping skills as automatically as AD clients resort to dissociation, withdrawal, and maladaptive covert and overt behavior. By honestly facing themselves, accepting their strengths and weaknesses, realizing that they do not have to be perfect, and by actively dealing with their problems, they internalize a sense of mastery, power, and self-esteem, which facilitates their social and emotional growth. This forms the foundation for an active meaningful lifestyle.

People who fail to effectively cope with and solve conflicts or resort to maladaptive behavior to do so are referred to as neurotic, AD, psychotic, rigid, regressed, disturbed, or as not integrated. They rely heavily upon maladaptive coping skills that allow the negative emotions underlying their problems to remain intact. They also rely heavily on defensive behaviors that only partially solve their conflicts, which in turn engender feelings of helplessness, inadequacy, and self-hatred. Often because of a strong and habitual reliance on dissociation, their overt and covert behavior is out of their control and perceived as not part of them. People who cannot cope may feel alien to themselves because of the high levels of anxiety that emerge when defenses are down. Their interpersonal relationships are usually impoverished because they resort to infantile methods of coping that are not easily tolerated by the more well-adjusted people with whom they must interact in order to have an enjoyable or fulfilled life.

Dissociation is an important adaptive and maladaptive defense mechanism employed by many AD clients and chronic A-Bs. Congruent with this is that most A-Bs exhibit marked conversion symptoms and compulsive features. This was recognized as early as the late 18th century (Gull, 1874; Laseque, 1973). Janet (1903) subdivided anorexia nervosa into obsessional or hysterical types.

Breger, et al., postulates that the dissociative process is the core defense mechanism present in all ADs. My observations support his view. Dissociation as the clients' major defense mechanism has to be abandoned in order for positive behavioral change to occur. Resistance to change is often encountered with clients who readily dissociate because they unconsciously and consciously blunt all mediating emotions and cognitions, therefore preventing them from being extinguished.

People resolve conflicts by two processes: 1) integration: taking an active part in solving problems, learning a variety of coping skills, and honestly confronting fears and feelings, or 2) taking a passive stance: dissociating, denying, or splitting off conflict-producing or anxiety-eliciting thoughts and actions from the self. In order for clients to develop coping skills, they must internalize the belief that what they do can make a difference in obtaining rewards, and that their attitudes and actions determine the quality of their existence. The internalization of the belief that "I am able to affect the world around me by my behavior for my betterment" is incongruent with what is often at the core of ADs: feelings of anxiety, being out of control, and helplessness.

The higher the levels of anxiety, the more clients tend to resort to the defensive or coping behavior that was reinforced in the past. Chronic AD clients resort to dissociation, passivity, compulsiveness, etc., which guard them from experiences which would enable them to learn adult coping skills. They cheat themselves from learning mature coping behaviors by relying on immature avoidance behavior. As a result, they may appear fixated or regressed at an early developmental stage because they automatically employ immature conflict reducing behaviors. Feedback from others who are critical of them often reinforces their feelings of helplessness, inadequacy, and low self-esteem, and may cause them to resort even more to their maladaptive coping strategies. Avoidance behaviors are the easiest and most temporary means of reducing negative emotions, and are, therefore, immediately reinforced. The immediate reinforcement effect (the closer the reinforcer is to the behavior, the more potent the reinforcer becomes) helps maintain these behaviors and causes them to become habitual.

More mature coping strategies in the long run prove more adaptive and rewarding, but are not immediately reinforcing. Therefore, many people choose the easiest way out, which is usually not the most adaptable. One would think that once clients realize the positive aspects of learning more adult coping skills they would make a strong effort to learn these skills. However, this is usually not the case. Again, reinforcement becomes significant in maintaining the maladaptive behavior; not only is the behavior reinforced by anxiety reduction, but it may also be reinforced by secondary gains. Also, most of us are simply lazy and take the easiest way out of a predicament.

It is understandable how clients can become victimized by their habitual use of dissociation as a defense against negative emotions, and resist abandoning it as a major coping strategy. This is especially true if it is the only coping strategy that was acceptable to the parents or other authority figures. Parents and authority figures who are rigid and overly rejecting punish the child's attempts at assertive coping behavior such as venting anger, questioning the wisdom of their attitudes, etc. Rigid authority figures restrict the child's choices of coping to only dissociation. Anger and other unacceptable but normal emotions and behaviors are then dissociated by the child. Dissociation is reinforced by authority figures' praise and various other rewards. I often heard parents describe their child as having been a "perfect child" before the onset of their disorder. "Perfect children" are usually so insecure that they do not dare openly venture into any activity where a conflict with the parents or others may result. Assertive action by them may be interpreted as hostile or asocial behavior by insecure parents or caretakers. Thus the parents and society have shaped the "perfect" externally-compliant child who is in actuality raging, rebelling, and frightened inside.

Loss of control occurs when people have dissociated thoughts, feelings or behaviors to the degree that these behaviors function outside the realm of what they consider the self. Dissociation of feelings, thoughts, and behavior increases the clients' already high level of anxiety by the addition of the fear of loss of control. The internalization of the belief that they are out of control adds to their sense of helplessness and low self-esteem. These feelings further deepen depression and maladaptive overt and covert coping behavior.

In order for people to master the developmental tasks encountered in society and to learn adequate conflict resolution skills, they must internalize a sense of security. Anxiety is the converse of this necessary security, and the prototype for this insecurity is separation anxiety (Breger, et al.).

Whether or not the dissociative or integrated stance is taken, as well as to what degree or at what point in the developmental sequence it occurs, is a function of the person's past experience with significant people and his environment. Hereditary predispositions also affect the degree to which a dissociative or an integrated stance is taken. The interaction between these variables determines the clients' basic level of anxiety and the severity of their maladaptive behavior. Reinforcement and punishment determine which coping strategy is employed; again, inherited predispositions may be a powerful variable. Dissociating and maintaining a facade of compliance and denying negative feelings is reinforced by most significant others. Seldom is an aggressive assertive act by a child rewarded by the adult caretakers of our society as much as passive compliance. In our culture, as well as in most others, girls particularly receive more reinforcement for dissociative coping behavior than boys. This partially explains why A-Bs, depressives, and clients with dissociative disorders are predominantly female.

I have observed that clients with psychophysiologic complaints such as colitis, hypertension or ulcers resort less to dissociation than those with dissociative disorders, OCD, or addictive-compulsions. Clients with psychophysiologic complaints are not dissociating their anxiety as well as dissociative clients, and instead feel it and suffer the psychophysiologic consequences. Generally clients whose major complaints are psychophysiologic less often exhibit severe psychopathology. These clients usually respond quickly to relaxation techniques and biofeedback vasodilation training. For example, I have treated many cases of irritable bowel syndrome with thermal biofeedback, improved diet, and hypnobehavioral therapy. Treatment in these cases usually takes less than ten hours.

The anxiety that mediates and serves as reinforcement for A-Bs, though usually based on separation anxiety, can result from a variety of learning experiences. As previously mentioned, these range from specific traumas where food had been paired with negative feelings, sexual molestation, or a sequence of auto-conditioned negative feelings during which food was paired with cognitions that elicited negative emotions. In most cases, the basic insecurities stem from the programming by society that 'unless a woman is thin she is worthless.'

Sullivan (1947) viewed compulsive behavior and perfectionism as a method of coping with feelings of insecurity and uncertainty that resulted from being reared in an unloving family. Salzman's (1980) view is similar in that he states that the major unfulfilled need in OCs is to be loved and accepted. In support of this position, I have found that in most A-Bs the extreme negative emotions observed during abreactive extinction and in-vivo extinction originated from feelings of rejection and loss of love from parents and significant others. Salzman's and Sullivan's views support Breger's hypothesis that the strongest negative emotion that primates can experience is separation anxiety, and that separation anxiety is the prototype for most later anxiety.

Whenever children internalize the idea that parental love and acceptance by peers is conditional and that their self-worth is contingent on their behavior, the seeds for OCD behavior and probably most ADs have been planted. Compulsive behavior can also result from modeling the behavior of parents, and, probably more importantly, their peer groups (Harris, 2009). The peer group is often more important than the parents in molding how children and adults think and behave. The most potent etiological variable is the acceptance of the idea by the client that 'I am only worthwhile as long as I'm successful (in school, work, sports, etc.); if I am not successful, I am worthless and should not and will not be loved.' Through hypnotic regressions, I have found most A-Bs as children did not feel parental love or feared that if they had 'bad thoughts' or failed at anything, the love would be withdrawn. They felt that if they were 'found out,' their parents' love would be withdrawn and they would be emotionally abandoned. Some children are so imaginative that even the thought of abandonment elicits strong anxiety and fear. Therefore, the child is forced into dissociating those negative thoughts and feelings. It must be strongly emphasized that many children are so hypersuggestible that they accept as truth their own incorrect conclusions about what they think others are thinking, and then guide their lives by those misguided conclusions.

The caretakers of many children are often deficient in parenting skills and are themselves compulsive, rigid, and unforgiving in their relationships. This coldness elicits fears of abandonment and anger in children. With A-Bs and chronically obese clients, food is used as a substitute for self-love and positive relationships. The painful emotions and thoughts are coped with by dissociating them and complying with authority figures' and peer groups' strict standards of behavior. This causes more external compliance and explains why A-Bs are so insecure that they believe they have to be a 'perfect' person to be valuable. When insecure people fail or even encounter a situation where they may fail, they experience extreme separation anxiety and attempt to initiate a variety of avoidance strategies.

Children tend to view and treat themselves as significant others and their peer group have viewed and treated them. Overly critical and rejecting parents, authority figures, and peers cause children to view themselves in the same manner. The basic insecurities that originated because of the perceived rejection(s) cause children and adults to fear their own actions and thoughts. These fears motivate them to internalize overly rigid standards of behavior, which cause them to dislike and reject parts of their personalities. This forms the foundation for the unconscious self-hating and destructive fragmented parts of the personality which I have observed in many chronic A-Bs and ADs. Self-hatred is usually present as a mediator before the overt maladaptive behavior and is a potent causative variable. Maladaptive behavior that is perceived as being out of character for the client causes further self-hatred and anxiety, which increases the desire to perform the maladaptive compulsive behavior. A self-destructive feedback loop now forms in which negative feelings mediate maladaptive behavior, and the maladaptive behavior in turn causes more negative feelings. A thorough description of these destructive feedback loops is presented in Chapter M, *"The Bulimic Cognitive-Behavioral Feedback Loop."*

Dreams, play, and fantasy are important in a child's development. Children resort to fantasy when they are angry, hurt, etc., and resolve their conflicts with significant others through its use. Children are seldom secure enough or feel powerful enough to vent their negative emotions in the presence of these people. When children vent their negative emotions openly, their emotionally troubled and rigid caretakers and peers usually deal too harshly with the behavior or belittle the child's efforts. This results in the children internalizing the belief that their behavior does not significantly alter their environment, thus leading to helplessness. These negative feelings lead to fantasized revenge against the significant others, which leads to even more insecurity because of the incongruity between the desire for revenge and their overly rigid self-concept. Children with rigid caretakers may believe that even having one 'bad thought' is an unpardonable sin, and their religious upbringing all too often reinforces that belief. They may be led to believe that God is mean and unloving and will send them to Hell for every little 'sin.' Statements by a fanatical religious coach such as, 'you are unworthy of God's love,' helped reinforce starving in a hypersuggestible 16-year-old athlete. The damaging effects of religious programming are described in Chapter N: Spiritual Aspects of Psychotherapy.

The resultant increase in anxiety adds to the children's high baseline anxiety and motivates them to dissociate the anger, hurt, and desire for revenge. In other words, the children dissociate or suppress these feelings and fantasies because of the separation anxiety that these feelings elicit. As previously stated, children make both correct and incorrect inferences about what parents and significant others are thinking. Incorrect and often negative inferences are uncritically accepted along with the valid ones and may become deeply ingrained in the children's unconscious to the point of guiding their behavior.

Uncritical acceptance of and rigid adherence to an inference or suggestion is characteristic of anxious adults, children, and somnambulistic (hypersuggestible) hypnotic subjects. The higher the anxiety, the more suggestible people become, and the more readily a suggestion may become implanted into their unconscious mind. Somnambulism and its relation to A-B and neuroses in general is discussed later in this chapter.

Clients' unadaptable ideas of how the world operates and their negative self-concept and image have to be changed and made congruent with their conscious wishes and goals. Unless overly-learned expectations and self-concepts are congruent with conscious goals, these variables will sabotage the actualization of those goals. The importance of self-image, expectations, or unconscious suggestions as directors and limiters of our personal growth has been recognized by many popular theorists (Maltz, 1960; Kroger & Fezler, 1976).

Negative autosuggestions and their unconscious effects can be identified and altered through hypnosis. The schemas, expectations, images and literal suggestions which have been accepted or modeled from significant others are the residues from childhood that live within the adult unconscious mind. Again, I must emphasize that I identify the unconscious mind as that part that contains the early learned CERs, etc., which cause us to respond automatically. These unconscious schema guide the adult behavior and represent the resistances that are encountered when attempting to enact positive behavioral change through psychotherapy. Often the resistances take the form of an alter-personality that is fixated at a particular stage of development. When clients become stressed, they automatically regress to this state and perform the compulsive behavior. Rapid and sometimes very obvious changes in states are characteristic of Dissociative Disorders, A-Bs, ADs, and OCDs, as well as when anyone is stressed.

In summary, overly anxious children experience an excessive amount of anxiety when encountering the developmental conflicts and situations that present the possibility of experiencing failure. When incapacitating, anxiety makes adaptive and assertive problem-solving unrewarding. Therefore, these children choose to withdraw and dissociate the various feelings by avoiding developmental tasks which are necessary for adequate development. A variety of early experiences, including separation anxiety, rejecting parents and caretakers, or a violent or harsh social environment, may cause them to feel that they cannot master the forces of the outside world. This causes them to resort to fantasy resolutions that, because of their vengeful and emotionally painful nature, cause more anxiety. High anxiety motivates children to dissociate negative emotions and avoid situations that need to be encountered in order to learn adaptive coping skills.

When children react to their own impulses and thoughts as cues of impending rejection, anxiety becomes internalized. Emotions elicited by forbidden thoughts and behavior are powerful. Often these emotions are of the same magnitude as those children would experience if their parents directly told them that they did not love them and were going to abandon them. Conditioning these negative emotions to covert processes (thoughts) is the mechanism of internalization. The result of this internalization is the development of a conscience - the internalization of a system of right and wrong. The development of a conscience is necessary for the child to become socialized, but when it is too strict and elicits strong guilt, it hastens dissociation and anxiety-mediated behavior.

I must again emphasize that children view themselves in the same way that they (correctly or incorrectly) infer that their caretakers and peers view them. Children's sense of what they consider self and non-self are determined by the aforementioned. Children's denial of their own humanness, i.e., feelings of anger, can cause the splitting of the personality and the formation of another part who acts outside the self.

Most A-Bs and compulsives are perfectionists who believe that living up to stringent expectations is the only way that they are valuable and worthwhile. This thinking is modeled from parents, peers, and the popular media. Most caretakers evaluate their own worth as well as others' by external standards of accomplishment and productivity. Insecure compulsive caretakers and peers increase children's insecurity by being overly critical and intolerant of children's mistakes. If children are never accepted as they are, they can conclude they will never will be good enough to be accepted.

They come to believe that being accepted and loved means achieving unreasonable goals. For example, I have encountered clients who hate themselves because they were only occasionally the best student or did not become an Olympic gold medalist. Suggestible children may incorrectly infer that their parents, peers, or others are demanding perfection when in reality they are not. Once an idea is accepted by a child and becomes overly learned, it may be hard to change when the child becomes an adult.

Compulsive perfectionistic behavior is reinforced by praise received from children's caretakers and peers. Of course, the praise is sporadic so the partial reinforcement effect solidly cements this maladaptive thinking. As children mature, a wider range of competitive situations and more skillful peers are encountered, and more opportunities for failure are encountered. These situations in turn elicit strong feelings of separation anxiety because failing means that they will not be loved and may be abandoned. This explains why much maladaptive behavior emerges around adolescence or college-age, as this is when more competitive situations are encountered.

Insecure hypersuggestible people who rely heavily on dissociation seek a simple solution for their problems and readily accept dichotomous thinking as a cognitive style. Everything in dichotomous thinking becomes orderly: there is a 'right' and a 'wrong' with no values in between. Dichotomous thinking is observed in most A-Bs, compulsives, and depressives. They perceive themselves to be either good or bad, successful or a failure, thin or fat, perfect or imperfect, on or off a diet, a non-drinker or a drunk. Any failure they experience is interpreted as a direct reflection of their self-worth and totally negates all previous accomplishments.

Trance-logic is the ability to accept and at the same time become unaware of logical incongruities. The ability to uncritically accept suggestions is characteristic of somnambulistic (the most suggestible) hypnotic subjects. Somnambulists can experience negative and positive hallucinations while under hypnosis. For example, a somnambulist can be given the suggestion that there is a clock on the wall when in reality there is none; when asked to give the time, he does so, just as if the clock were really there. Obviously non-psychotic people who can produce hallucinations have a powerful imagination and profound ability to suspend critical thinking. Somnambulists, who naturally employ dissociation as a primary defense mechanism, are usually superior hypnotic subjects and habitually employ trance-logic in their daily lives. Somnambulists not only have the ability to suspend critical judgement, but to also telescope time so that they are focused only on the present and therefore unable to relate past experiences to their current experience. They also exhibit a sluggishness in objectively scrutinizing situations and lack the ability to objectively view their own behavior. Many chronic A-Bs, ADs and OCDs exhibit these characteristics. An A-B's trance-logic solution to all her problems is: 'If I can lose weight and look like a fashion model, all my problems will be solved. I can never be too thin because losing weight is always good. It is the measure of my value. Only thin girls are, or should be, loved.'

When the A-B embarks on a diet, she suggest to herself that she is worthwhile when she is adhering to it perfectly and worthless when she is not. Slipping off the diet elicits extreme anxiety (separation anxiety). Breaking the diet is viewed as an irreversible catastrophe and causes the A-B to experience guilt and may motivate self-punishment. The typical thought is: "I failed, so I may as well 'pig-out.'" This leads to binge-eating and vomiting. There is no happy medium for an OCD perfectionist. The internalization of the 'all or none' orientation explains the extreme mood fluctuations that A-Bs, depressives and most OCDs experience.

Compulsive-perfectionists become trapped in a vicious cycle. Their unreasonably high standards ensures that they experience more failure than success and therefore more self-hatred than self-love. The compulsive drive to be perfect causes them to emphasize and become focused on important as well as unimportant details. This may cause them to not complete the important tasks that would be rewarding and that would help prevent depression because they become waylaid by insignificant details. Any flaw in their work, no matter how insignificant, is inconsistent with their rigid self-concept; therefore it cannot be tolerated. To make a mistake or to admit to a mistake is perceived as a direct attack upon their already shaky self-concept and is avoided at all costs. This causes them to be overly careful and to expend time and energy needlessly, increasing the probability of falling short of their goals. Falling short of their goals exacerbates their self-hatred and depression, prompting a vicious feedback loop begins.

Extreme perfectionism leads to difficulties with employers, loved ones, and peers. Being admonished for their rigid behavior increases their sense of low self-esteem and insecurity and often causes them to resort to attempting to become more 'perfect,' as this is the only way they know to feel worthwhile. Losing a job or dropping out of school in order to avoid anxiety along with causing self-hatred eliminates many of the reinforcing activities that made their life at least somewhat enjoyable. Usually perfectionists have put so much value on these activities that they end up with few if any other rewarding activities to rely on. The loss of these reinforcers along with the resultant inactivity leads to a sense of helplessness and depression. The term "joyless overachiever" characterizes many people in the Western World.

The following is a typical example: The client was a 17-year-old female high school student who transferred to a prestigious eastern prep school. In spite of her understanding that she had been grouped with the best students in the country, she nevertheless had great difficulty in accepting any grade less than an 'A' or not being number one at everything. Whenever her performance put her in the middle of the pack, she experienced great frustration, anger, and depression. She reduced these emotions by dissociation, binge-eating and vomiting. She was unprepared for coping with failure or even minor setbacks because she had never encountered defeat.

The lessons from experiencing defeats and overcoming them through hard work and determination are necessary as it extinguishes fears of failure and gives individuals a realistic sense of self efficacy. Additionally, a person can learn to experience great joy from overcoming obstacles through hard work instead of quitting the battle. This client's fear of failure along with her extreme perfectionism mediated and reinforced her self-hatred and maladaptive behavior. She had failed to acquire the important attitude that the most important thing in life is not to be liked or admired by others, although this is certainly enjoyable, but to love and unconditionally accept one's self. She must learn to appreciate that she is a unique individual with infinite value, and that this is independent from her grades or other accomplishments.

The acceptance of these attitudes or beliefs is the basic goal of many psychotherapies and form the core for adaptive mental adjustment. I have never treated anyone for reactive depression or A-B who had an internalized sense of self-love and acceptance of themselves as imperfect but possessing value. These values and beliefs should be ingrained into children's minds at an early age through their religious upbringing and by the caretakers' actions and literal statements. School grades and other achievements will then take their proper place as second to life and health.

When working with clients who are serious Christians, the Biblical teachings concerning self-love, God's love for man, and the Christian doctrine concerning guilt should be explained. For an explanation of how to incorporate the Biblical teachings into therapy, the reader is referred to Chapter N: *Spiritual Aspects of Psychotherapy*. The majority of my clients have professed a belief in God and discover that much of their guilt and fear have to do with their religious upbringing. Many MHPs do not take the clients' spiritual beliefs seriously and are often unprepared to deal with spiritual conflicts.

Most perfectionists perceive themselves as ineffectual and unsuccessful because they believe that successful people achieve their goals with total self-confidence, minimal effort, few errors, and no emotional distress. Because of this misconception, perfectionists view their own coping behaviors, anxieties and setbacks as weaknesses that others do not experience. This erroneous thinking leads to a further lowering of self-esteem, more self-hatred, and withdrawal (dissociation) instead of actively working at overcoming the obstacles that need to be overcome. Their motivation is weakened by the expectation that they will fail or that being successful would involve so much pain that it is not worth the effort. All of the above reinforces their feelings of helplessness and strengthens their beliefs that they do not deserve to be successful and that success is for others and not for them. Perfectionists will come a long way to achieving better adjustment when they accept that most successful people do not have fewer problems than others, but instead learn from their failures and continue striving instead of giving up.

Many chronic A-Bs, Ads, and depressives are deficient in social skills, and fear of rejection prevents them from forming rewarding interpersonal relationships. They are therefore often rejected by others and plagued with loneliness. Rejection increases their insecurities and mediates more maladaptive coping behavior. Most A-Bs are overly concerned with other people's opinion of them and will go to great extremes in order to be accepted. Losing weight and becoming more perfect is of course the A-B's way of becoming more acceptable.

Compulsives, because of their high levels of anxiety, become rigid, intolerant and overly sensitive to what they interpret as criticism. These defensive behaviors often frustrate and alienate others, who then reject them. This rejection may lead the compulsive to conclude that, 'I must become more perfect in order to be accepted,' or to rely upon the trance-logic conclusion that, 'I was rejected because of how I look.' The real reasons for their problems are not clearly understood because they are unable or reluctant to objectively evaluate and observe their own behavior. Most emotionally healthy people experience this difficulty. Simple self-confrontation by facing one's self in a mirror or listening to one's voice has been shown to be unpleasant (Sackeim & Gur, 1978). These unpleasant emotions often mediate the avoidance of encountering the self and objectively facing one's problems.

Perfectionists and defensive clients may become reluctant to communicate with the therapist because they fear the therapist will reject them. Therefore they rationalize and lie in order to make themselves look good. Fear of rejection often causes them to mobilize a set of defenses that exclude the people who would be most able to help them. Many clients feel that they cannot trust anyone. This basic mistrust is usually the product of being raised by rigid-perfectionistic caretakers who could not be trusted to unconditionally love and accept them.

As perfectionistic-insecure children mature, the amount of reward for perfectionism begins to diminish. They find it increasingly difficult to live up to significant others' strict standards. As they ascend through the educational system, they encounter more and more competitive situations and the gap between their expectations of being number one and their actual performance widens. This of course causes more anxiety and insecurity, which they again seek to reduce through dissociation, compulsive ritualistic behavior, and starvation and/or binge-eating and vomiting. The fears of failure may become so strong that they may withdraw from all competitive situations. As a graduate student, I saw many who became paralyzed and never completed their dissertations due to their anxieties. In all walks of life there are many who feel they have to 'win' all the the time and, because this is impossible, end up avoiding all competitive situations.

As previously stated, the lack of an internalized sense of realistic self-efficacy is the basis of these maladaptive behaviors and cognitions. Instilling in clients a spiritual sense of self-love and an understanding of the importance of self-love allows them to better assess the importance of grades, being number one, and so forth. Self-love and health must become more important than being perfect and winning.   Many people have led wonderful fulfilling lives without being anywhere near number one at anything. Instead, they experienced joy in their life through finding meaning and satisfaction in their work and relationships. Situations which involve competition and the possibility of failure then become less anxiety-eliciting because they do not mean loss of love or self-esteem. When winning becomes less important, it allows the person to perceive competitive situations as adventurous challenges and these situations then become more fun.

When competitive situations are approached in this manner, the person's chance for success is increased because a lower level of anxiety usually helps a person think more clearly and perform better. The perfectionists' standard for evaluating the meaning in their existence is based upon external and therefore often uncontrollable variables. This leaves them chronically insecure because these externals can be taken away. An important therapeutic goal is to help clients internalize an attitude of active participation in life along with a feeling of self love. This is done via hypnotic suggestion and education. Also, the behavioral treatment for depression and the treatment for OCD outlined in Chapter H (*Cognitive-Behavioral Treatment of Depression*) force clients into reinforcing activities and cause many of their anxieties to extinguish.

A strong caution must be given concerning self-esteem. Many MHPs and teachers have become deluded and accepted as an absolute truth that enhancing people's self-esteem is essential for them to become good human beings. I must emphasize that a realistic sense of self-esteem and the implantation of a moral code is essential. Countless studies have shown that juvenile delinquents and other destructive people possess higher self-esteem than people who function as productive members of society.

As a boxing coach, I have been around criminals all my life, ranging from street-muggers to more professional criminals. At one time, four of my boxers were in prison for murder, and between them, they had killed at least six people. Not one had low self-esteem; in fact they all felt good about their predatory skills and their lives, except for being in prison. When I was younger, I also had contact with a group of career criminals who thought they were 'special' and therefore invincible. At the time they drove fancy cars, dressed well, and had attractive girlfriends. They felt good about themselves as well since they had acquired certain material things and "respect" from their peers (although they hated prison). All in all, many criminals are narcissists and usually hold themselves in high esteem, as they believe they are smarter than others. They honestly believe the rules that apply to the rest of society do not apply to them.

A realistic sense of self-esteem must be accepted by an individual in order to have a successful life. Many MHPs and teachers, however, have cultivated an unrealistic sense of entitlement and the idea that everyone is a winner; these ideas must be dispelled. The definition of success should be spelled out to children and clients. People are successful each day they love themselves, feel good about who and what they stand for, and are progressing toward a worthy goal. This definition allows for failures and setbacks because they are perceived as necessary challenges that are to be overcome as part of the pursuit of worthy goals. Successful people are successful problem solvers and do not expect to get something for nothing. All they ask for is to be able to compete fairly on an even playing ground.

When insecure children reach late adolescence and are supposed to leave home, they are suddenly expected to become independent and socially skilled adults. As previously mentioned, this prodding by parents, peers and society explains why A-B and other problems emerge at this time. These adolescents were cheated from encountering the painful experiences that, when mastered, would build self confidence. Taking the easy way out is seldom the most beneficial from the standpoint of learning new coping and life-enhancing skills. It is easier to lie and play a false role, especially when this is being reinforced by caretakers' acceptance, than to encounter the world in a direct way.

Dissociation is common to most chronic ADs and so must be thoroughly understood by both clients and therapists. As previously stated, thoughts, feelings, urges, etc., that are incompatible with the person's self-concept and conscience are often dissociated from the self.

Thoughts and feelings that elicit anxiety cause children to resort to dissociation of these feelings and thoughts, and to relegate them to another part of the personality. Forbidden thoughts and emotions are dissociated and not considered part of the self but part of 'the other one.' This reaches its extreme in Dissociative Identity Disorder. Clients' actions and ideas are treated as if they are not their own, and the dissociated actions occur while they are in a trance state. During the trance state, time is distorted and complete amnesia may occur. Clients may also report that they feel like an outsider observing their own actions and that their actions do not feel like part of them. All emotions may be blunted during this state, and this decrease in anxiety reinforces them entering the trance whenever negative feelings are encountered.

Dissociative or trance states, like most behaviors, can be controlled by either external or internal stimuli and, through repetition, may become an automatic coping strategy. This explains why many bulimics state that they binge and vomit automatically and are aware their binge-eating and vomiting is a means of gaining oral gratification while avoiding general tensions or specific negative emotions. Bulimics often report that time passes very quickly when they are bingeing and purging; some also report partial to almost complete amnesia. Entering a dissociated state during which a person works for hours on a creative project is experienced by many. Normal mentally well-adjusted people do this and report how quickly time passes. Bulimics simply use this creative dissociation ability in a maladaptive way.

In order to extinguish the automatic trance, the eliciting stimuli have to be identified. These stimuli are repeatedly presented causing repeated elicitation and extinction of the emotions which trigger the trance behavior. This is similar to FRP, except that the behavior being extinguished is entering a trance rather than a compulsive behavior. Eliciting the emotions that mediate the withdrawal into trance through flooding with the relevant stimuli (i.e., high calorie food) and preventing the client from entering a trance in order to blunt these emotions allows extinction of those emotions to occur.

Abreactive extinction through hypnotic regression can decrease the baseline and specific anxiety that mediates trance behavior. As the mediating emotions and baseline anxiety extinguish through FRP and/or abreactive extinction, clients automatically escape less often into trances. Since much of the binge-eating and vomiting occurs during trance states, this behavior also diminishes. As the anxiety and trance behavior decrease, clients' hypersuggestibility usually diminishes and they resort less to trance coping. They are then free to use their trance behavior for creative endeavors.

Observations of thousands of clients by Kappas (1978) support my finding that, once negative mediating emotions are extinguished, hypersuggestibility and the frequent entering into spontaneous trances decreases. Kappas, et al., states that anxiety and suggestibility are positively correlated, and my experience generally supports his conclusion. Decreases in suggestibility and anxiety usually parallel clients resorting less to dissociation and to more actively resolving their conflicts by applying the adaptive behavioral methods they have learned.

The idea that anxiety and suggestibility are correlated, however, is not supported by the research literature (Hilgard, 1965). On the other hand, one need only observe stage hypnotists using confusion induction techniques and revivalist ministers performing religious conversions and healings to see how increasing confusion and anxiety often increases suggestibility. For example, revivalist ministers increase the emotionality (i.e., fear, anxiety, guilt) of their listeners by convincing them that unless they accept Jesus as their Saviour, they will burn in Hell. Following this constant barrage, members of the audience enter trance states. The minister asks for those who are ready to be 'saved' to come forward. The ones who do are those who are in a trance (hypersuggestible state) and have already accepted most of the suggestions given by the minister and ready for the rest of the message to be implanted. When their turn comes, the minister makes a declaration while applying a sudden physical shock (such as a slap on the forehead) which drives the person into a state of hypnosis, often characterized by the eyes rolling up and the entire body shaking. The religious conversion, like a hypnotic suggestion, may last for years and perhaps a lifetime even if it is only occasionally reinforced. Some individuals in the audience may also enter hypersuggestible states and be healed of a dissociatively-based malady without a formal religious conversion simply because of the emotionality of the group event.

Some clinicians employ rapid induction techniques that increase confusion and anxiety in order to induce hypnosis in already tense subjects. Kappas, et al., has reported decreased skin resistance in many subjects, which indicates anxiety (arousal) precedes the subject entering the trance. Apparently increased suggestibility as a result of anxiety allows rapid learning of simple avoidance responses and fear. It is adaptable for an organism to be hypersuggestible during a state of anxiety or fear so that it more readily acquires adaptive avoidance behaviors. It is adaptable for organisms to learn and never forget behavior which allows successful escape from or avoidance of dangerous situations. Therefore, relaxation and positive states incompatible with fear are also very powerful. The joy of becoming one with Jesus and the expectation of unconditional acceptance and God's love explains the power of religious conversions.

Heightened anxiety can open the way to dissociative or hypnotic states and people are at their peak of suggestibility during these states. Low states of arousal also probably correlate to hypersuggestibility (Budzynski, 1976), which explains why a progressive relaxation induction also can initiate hypnosis. Budzynski, et al., also does not exclude the possibility that high states of arousal also correlate to suggestibility. Little research has been done concerning this topic because increasing anxiety and fear obviously invokes ethical considerations.

The cognitive and behavioral characteristics of hypnotized subjects, ADs, ABs, and OCDs are often similar. Both hypnotized and non-hypnotized people who are in a high anxiety state experience heightened suggestibility, time distortion, and a suspension of critical thinking. For example, OCDs become totally and uncritically obsessed with one set of thoughts and lose the ability to objectively scrutinize their behavior. Focused selective attention is also characteristic of a deeply hypnotized person. An extended discussion of this topic is included in Chapter L: *Hypnobehavioral Models of Obsessive-Compulsive Behavior and the Unconscious*.

I have encountered a few clients in which numerous FRPs in combination with abreactive extinction have failed. One client unconsciously prevented the extinction of the emotions mediating her extreme binge-eating and vomiting and starvation during seven consecutive days of FRP. The client was hypnotically regressed and a fragmented part of the personality which was responsible for the resistance was brought into consciousness. The regressed part of the personality was determined to punish her mother by killing the adult part of herself. Through hypnotic suggestion and separation therapy (discussed in Chapter K), her unconscious motives changed. However, because she had received little benefit from the extinction trials, the FRP had to be done over again. It is interesting that the client's dissociated motives were able to prevent extinction from taking place. In this case, she was able to successfully extinguish most of her fears and negative mediating emotions on her own in her home environment. This woman dissociated so much of herself that she created another 'self' that, when questioned, did not realize she would also die if she were successful in killing the 'adult self.'

Breger, et al., describes the integration: "To become "conscious" of a dissociate complex is to own up to it - to comprehend its meaning as part of one's self" (p. 205). I am basically in agreement with Breger's statement. However, his statements concerning bringing the dissociated complex into consciousness may lead one to incorrectly assume that adaptive behavioral change will automatically follow. Causing the client to become conscious of the dissociated complex seldom alone induces desired sustained behavioral change in chronic A-Bs, ADs, and OCDs. Insight without the clients' acceptance of important cognitions such as an increased self-love and the attitude that they can overcome their problems by facing them will usually not result in behavioral change. Also, if these cognitive changes are not reinforced by practice, lasting behavioral change will usually not occur. However, psychotherapy (including hypnobehavioral therapy) never occurs in a vacuum, and making dissociated motives conscious through separation and trance therapy may result in striking temporary behavioral changes.

Integration (having the parts act in harmony) does at times initiate behavioral change. Often the maladaptive behavior immediately subsides while clients recognize and keep separate the dissociated parts of the personality which subserve that behavior (Kirsten & Robertiello, 1975). Whether this is done through hypnosis or imagery, the results are the same. Hevesi's trance therapy probably works because of the aforementioned separation and integration processes and the positive suggestions that he gives while the client is in a suggestible hypnotic state. Both separation therapy and trance therapy are described in Chapter K.

In summary, behavioral theorists' strict criticism of the benefits of insight and cognitive change and unconscious variables is not entirely justified. Unconscious motives, dissociations, attitudes, expectations, and suggestion do influence the extinction process. Causing clients to become aware of fragmented parts can help eliminate unconscious resistances that prevent them from deriving benefit from the extinction procedures. However, behavioral therapists are correct in their criticism of having clients simply talk about their problems or continuing to probe the past to have the clients gain insight into why they are suffering. MHPs that repeatedly spend many hours searching for every trauma such as sexual molestation, etc., end up finding false memories of traumas. After months and hundreds of hours of this, the clients become more confused than ever (Loftus & Davis, 2006).

Even when there is trauma such as sexual molestation that is the CER causing problems, I only elicit it to extinguish it. Eliminating suffering is important, not gaining insight or determining how accurate the memories are. Buddhism is congruent with the behavioral approach. Many MHPs keep focusing on the past rather than on the 'now.' They believe that understanding of one's suffering will help relieve it. As Shakyamuni, a Buddhist, once said, "If one comes across a person who has been shot by an arrow, one does not spend time wondering about where the arrow came from or the caste of the individual who shot it or analyzing what type of wood the shaft is made of or the manner in which the arowhead was fashioned. Rather, one should focus on immediately pulling out the arrow" (Dalai Lama & Cutler, 1998).

Hypnotic suggestion and spontaneous changes (which can often be beneficial) that occur during trance also greatly influence behavior. In some clients, repetitious direct hypnotic suggestion can build enough active inhibition to prevent spontaneous recovery. However, I along with others (Meares, 1972), have found that literal and direct hypnotic suggestions given to A-Bs in the attempt to have them feel hunger and eat are counterproductive.

Meares, et al., treated a seriously emaciated 18-year-old anorexic girl with post-hypnotic suggestions to increase her appetite and eat a good meal that night. Over a few days she became more anxious and lost weight. Her regressed behavior also became more pronounced, and she dressed and spoke like a ten-to-twelve-year-old child. It is reasonable that post-hypnotic suggestions to engage in a highly anxiety-eliciting behavior would cause heightened anxiety. This heightened anxiety is the mediator for A-Bs' dissociative defense reactions and starving. Basically, A-Bs vigorously enact their overly-learned and temporarily reinforcing defensive mechanisms of dissociation and starving when encountering increased anxiety.

In summary, integration of the A-Bs' fragmented personality is facilitated by the extinction of the negative mediating CERs that maintained the dissociation. Also, the unconscious motives may need to be understood by all parts of the personality in order for extinction to occur. If clients are allowed to dissociate the negative mediating CERs, extinction will be prevented and the emotions will be only temporarily suppressed, if at all.

It is fascinating to view a personality integrating as extinction progresses. Clients often report without any prior suggestions that they are beginning to feel 'whole' and a deep sense of peace. Simultaneously, they experience a decrease in their hypnotic responsiveness and hypersuggestibility that apparently results from the decrease in anxiety. Since this is a correlational relationship, a causal connection between the decrease in anxiety and the decrease in suggestibility has not been conclusively established; however the phenomena are reliable.

Hilgard (1965) and other researchers have concluded that high levels of suggestibility can be demonstrated in well-adjusted subjects, and there is no correlation between psychopathology and hypnotic suggestibility. Bernheim (1887) and the Nancy School were probably correct in maintaining that everyone is suggestible under the proper conditions. Simply because people are good hypnotic subjects (suggestible), does not mean they are prone to A-Ds. However, the Nancy School and many later researchers did not reject the view that intensified suggestibility was a characteristic of dissociative reactions. Suggestibility as a leading symptom in hysteria was accepted by many authors, including McDougall (1911), Hirschlaff (1919), Satau (1923), Bleuler (1924), Morton (1936), and Fischer (1937). I have found that hypersuggestibility and somnambulism are highly correlated with clients' use of dissociation and the severity of their maladaptive behavior and resistance to therapy.

Extreme uncritical hypersuggestibility, where a person is likely to accept almost any suggestion, is obviously undesirable. Most modern researchers (Hilgard, et al.) have concluded that hypnotizability, trance capacity, or suggestibility is a stable trait and shows little variation throughout a person's life. These researchers recognize of course that certain variables such as age cause general trends in suggestibility, but that these trends are consistent within subjects although variable between subjects.

It has been my and Kappas' experience, however, that suggestibility can change remarkably during the course of hypnotherapy. Kappas, et al., also observed drastic reductions in suggestibility in a variety of clients following the abreactive extinction of the anxiety that was related to the origin and maintenance of their maladaptive behavior. He termed his technique 'circle therapy,' which is the same as abreactive extinction.

Hypnosis can be defined as a state of relaxed focused concentration characterized by increased suggestibility. Hypnotic procedures and related trance phenomena are useful in a variety of ways. Hypnosis and deep relaxation induced through imagery appear to lower the threshold for the emergence into consciousness of troublesome dissociated emotions. The emergence of these stimuli aids the therapist in identifying the eliciting stimuli so they can be used in the extinction procedures.

Subjects can be age-regressed or can pseudo-revivify the time, place, thoughts, emotions, and stimulus situations where the maladaptive CERs were learned. Pseudo-revivification (reliving the past events in the framework of the present time) is usually adequate to elicit the relevant CERs and cause extinction. Revivifications (subjects feeling that they are actually reliving the past event) are helpful, but not necessary. Again the mediating and reinforcing CERs must be repeatedly elicited and avoidance prevented in order for their extinction to become permanent.

As stated, many chronic clients are in a sensitized state and open to suggestions as a result of high levels of anxiety. Therefore, I must emphasize that similar to hypnotized subjects, they will only be responsive (hypersuggestible) to certain contexts, emotional states, and certain people. They usually are hypersuggestible to rejection, their rigid caretakers and/or peers, and/or other variables, but not to the MHP. This explains why clients may not respond well to therapeutic suggestions, although they appear to be good hypnotic subjects. Somehow MHPs must increase clients' suggestibility to them. This may be what therapists mean when they talk about establishing a therapeutic relationship with clients. Forming a trusting caring relationship with another increases their suggestibility.

Once the dissociated eliciting stimuli are identified and the relevant suggestions (cognitions) and CERs are elicited, clients can identify these CERs and cognitions as the ones that are mediating their maladaptive behavior. Through understanding the learning situations which underlie their maladaptive attitudes and behavior, clients become more motivated and less apprehensive as their behavior makes sense. This understanding increases rapport between the client and therapist and decreases clients' fears that they are losing their mind. Clients' clearer understanding of the 'hows' and 'whys' of their behavior increase their faith in the therapeutic procedures and their expectations of success. They also become more suggestible to the therapist's suggestions. Clients' increased expectations of success enhances their motivation to comply with the extinction procedures, even though these procedures are emotionally painful.

The relationship between the MHP and the client is of paramount importance for successful therapy. Clients must trust the therapist and understand the therapeutic methods. Throughout therapy, the MHP should explain the various experiences that the client will encounter. When the therapist can predict what the client will experience and can explain the client's subjective feelings by reference to specific laws of behavior, the client's faith in the MHP and therapeutic methods increases. This improves rapport and further increases the client's suggestibility and compliance.

Connecting the past with the present in terms of memories, symptoms and feelings by regressing the client to the time when the mediating CERs or maladaptive cognitions were acquired is similar to what Watkins (1972) termed an *affect bridge*. An affect bridge is formed by first intensifying the relevant troublesome emotions through hypnotic suggestion or imagery, and then regressing clients to previous events where they experienced these emotions. The troublesome CERs and cognitions are then related to the clients' current feelings. A symptom bridge can be formed between a present symptom and its origin by using the same method. The emotions and symptoms, such as conversion reactions, are then repeatedly abreacted and extinguished. Unless the client employs dissociation to fake the whole procedure or there are strong secondary gains, extinction will occur.

The purpose of age regression is to explore clients' past and relate it to the present situations to help them master the negative CERs and thoughts that were acquired and currently mediating their maladaptive behavior. Hypermnesia, which often characterizes hypnosis, is useful because it facilitates recall and the subjective sense of 'realness' of the imagined or suggested eliciting stimulus situations. The closer the imagined or suggested situation is to the original and current eliciting stimuli, the greater the generalization of the extinction process to the real environment. Free association and other non-hypnobehavioral methods are much slower and less reliable therapeutic methods as they only partially elicit the mediating CERs, resulting in only partial extinction.

Again, I must emphasize that hypnosis is not a reliable method of obtaining the objective reality of past traumas, etc., and has no place as a forensic tool. The hypermnesia that clients experience is not a film of the past and often is inaccurate. Accuracy is not important in therapy; it is what is in the person's mind and the CERs that are important.

As stated, the insights clients obtain through hypnotic regression are often beneficial when the resistances of clients need to be understood. At times clients spontaneously contact a dissociated part that is resisting getting better. Becoming aware of any dissociated part(s) that are behind the resistances helps clients to face the resistance and extinguish it.

The abreactive extinction procedure is usually fatiguing, but clients must understand that the higher the intensity of the emotions, the more complete the extinction. As clients feel the CER, conversion reactions are often elicited that have been used to avoid the mediating emotions and/or as a method of satisfying the clients' unconscious or conscious motives. The therapist should not stop the abreactive process when conversion reactions occur, but instead explain to clients the origin of the defensive maneuvers and how they are using these reactions to avoid facing the negative emotions.

Following the explanation, the therapist must teach clients how to extinguish them in-vivo and through imagery. Also, having clients employ the law of reverse effect is helpful. Clients are instructed to try as hard as they can to use their willpower to make the conversion feelings worse. As they try, the feelings decrease and clients realize that by using a procedure they can reduce their reactions that have been bothering them. This gives them a realistic sense of control. Self-induced flooding while in hypnosis and in-vivo flooding has proven to be very effective. The conversion reactions should be repeatedly elicited by suggesting the CERs that cause them, until the conversion reactions and emotions extinguish. If the therapist allowed these resistances (conversion reactions) to stop the abreactions, they would be reinforced. The conversion reactions of swelling feelings in different parts of the body such as the abdomen or thighs are obviously produced to block eating. Many A-Bs, if they touch food to their lips while in front of a mirror, report they literally see themselves getting fatter. It should be emphasized that it is physically impossible for someone to gain weight by simply touching a high-calorie food to their lips.

Sometimes the extreme sympathetic activity that occurs during an abreaction can trigger psychophysiologic reactions such as colonic spasms and tension and migraine headaches. The best way to handle these reactions is to demonstrate to the client how these reactions can be triggered through suggestion and imagining the stimuli which elicit the mediating emotions. I have clients trigger the headache and then have them change their thinking or imaginings to a set of stimuli that elicit relaxation (parasympathetic activity). Images causing vasodilation of the arterial tree, such as a warm sunny day on a beach with a cool breeze across the face (which causes constriction of the cranial and retinal arteries) facilitate relaxation. These images reliably increase hand temperatures and therefore arterial dilation (Maslach, Marshall, & Zimbardo, 1972).

One has to be careful with psychophysiologic migraines to not allow the aura or head pain to last too long. Once the relevant neurochemicals are released, the reactive vasodilation in the retinal and cranial arteries may not be reversible by suggestion; thus a full-blown migraine attack may ensue. Following the cessation of the head discomfort or symptoms such as photophobia and other visual or visceral sensations, clients are made to understand that since their imagination or suggestion caused the headache and conversion reactions, their imagination or suggestion can take them away. Clients now realize that they have control over these reactions. The same methods can be used to decrease most psychophysiologic reactions. If, following the abreactive procedure, a client is still experiencing psychophysiologic or emotional discomfort, it usually indicates that the CERs need to be abreacted and extinguished as extinction has not been carried far enough.

As mentioned, fear of loss of control is a major source of anxiety in A-Bs and probably all ADs and OCDs. Therefore during abreactive extinction, fears of loss of control are suggested along with the other mediating emotions (urges to binge and fears of weight gain) so that these fears will also extinguish. Eliciting fears of loss of control more closely emulates the actual stimulus situations that trigger the maladaptive compulsive behavior and therefore enhances extinction and the generalization of extinction to the real environment. Fear of loss of control, like any stimulus situation, can act as a stimulus to trigger maladaptive behavior and dissociation. Extinguishing as many components of the mediating CERs as possible reduces the generalization decrement that is expected when clients return to their normal environment. Also, resistance (feeling resistant) can be abreacted and extinguished in the same manner. Many resistant clients report that they can feel themselves blocking the MHP's positive suggestions. The stubborn resistant feeling is often based upon distrust or the fear that if the therapist gets to really know them, they will be rejected. The fears of rejection and so forth that subserve the resistant feelings are also extinguishable.

Kappas' ideas (1978) concerning somnambulism and his methods for decreasing suggestibility are useful in the treatment of clients who are overly suggestible. This is in agreement with Spiegel and Spiegel (1978) who state that somnambulists are already in hypnosis and use their profound trance capacity against themselves. Although they further state that hypnosis should not be extensively used with somnambulists, my and Kappas' experiences indicates that the trance state is the most powerful tool we have for treating these clients.

Kappas, et al., approaches somnambulists by putting them into a deeper state of hypnosis than they currently are in, and then decreases their suggestibility by what he terms circle therapy. As mentioned, circle therapy is actually abreactive extinction, followed by desensitization with relaxing imagery (imagery of the client being relaxed and performing the feared behavior easily).

Kappas, et al., reports that somnambulists become less suggestible and more receptive to therapy following abreactive extinction. One may disagree with Kappas' basic supposition that hypnosis and anxiety are the same; however, his methods are useful and should be given serious consideration.

K. Trance and Separation Therapy

I have observed that dissociative identity disorders and other dissociative reactions such as fugue states, amnesia, and somnambulism are more common than most current writers lead us to believe. More dissociative reactions were reported in the 1800's than today, but I can find no reason to conclude that there are necessarily fewer dissociative reactions occurring today. Changing patterns of maladaptive coping behavior (symptoms) and more diffuse and vague conversion reactions in combination with the movement away from using hypnosis may explain why dissociative states are often not detected. Classic conversion reactions such as glove anesthesia or hysterical blindness may be rare today, especially among more educated clients. Today's clients are more knowledgeable and better able to fool themselves and clinicians by producing symptoms that are vague and could have either an organic or functional basis. Such symptoms may include diffuse abdominal pain or body aches, headaches, back pain, nausea, allergic reactions and pseudo-seizures. Unfortunately, these symptoms may also indicate an underlying physiological illness that is hard to diagnose.

In support of the hypothesis that conversion reactions and dissociative phenomena are not uncommon is the finding that many women who are obese or pregnant, in addition to many who are of normal weight, consistently overestimate their body size. The common factor appears to be a high level of concern about their body size and shape (Button, Fransella, & Slade, 1977). Slade and Russell (1973) also demonstrated that anorexics in general considerably overestimate their body size.

On the other hand, all too often clients are diagnosed as having conversion reactions as a last resort when medical tests prove inconclusive. Older research showed that at least 60% of patients classified as experiencing conversion reactions had real organic diseases that were diagnosed later (Slater & Glighero, 1965; Whitlock, 1957). However, a review of recent literature shows the opposite.

It is, therefore, difficult to be sure whether or not there is an organic problem, a conversion reaction, or a combination when clients have been previously diagnosed as manifesting conversion symptoms. The diseases that usually are confused with conversion reactions involve the central nervous system (Slater, et al.; Whitlock, et al.). Dissociative states, such as fugue states, selective amnesia, or Dissociative Identity Disorder, are usually more easily identified and less often confused with organic malfunction. If the conversion pain or feeling can be moved to different areas of the body through hypnotic imagery or suggestion or if it appears during abreactive extinction, one can reasonably conclude that emotions are at least an exacerbating variable. The extreme body distortions that are experienced by A-Bs following the ingestion of a small amount of 'forbidden' food are obviously conversion reactions and easily manipulated by imagery and suggestion.

Many modern psychologists, especially those trained in non-directive approaches, never ask questions such as, "Do you experience periods of time when you do things but then not remember what you have done?", or simply, "Do you experience time distortions?", that would determine whether or not a client experiences dissociative states. Some clients report that occasionally after a few drinks, which would not normally affect them, they found themselves saying and doing things that were incongruent with their usual behavior. These clients may be using alcohol to act out dissociated motives. Rapid changes in mood may also indicate that the client is experiencing mild or severe personality state changes.

Dissociative reactions are not an all-or-none phenomenon, but occur on a continuum from mild to severe. There are probably few true Dissociative Identity Disorders. However, it is reasonable to assume that milder fragmentations of the personality are more common. Many A-Bs, OCDs, and ADs report that they feel fragmented and state that, "One part of me wants to do one thing and another part of me wants to do the opposite." An obvious sign that most A-Bs experience dissociative states is that they experience time distortion or partial amnesia during binge-eating and vomiting. Some A-Bs experience complete amnesia for whole blocks of time and may manifest regressed and infantile behavior when stressed. This is usually most evident during the bulimic episodes. Anorexics also experience similar reactions although their reactions may be less obvious. Some experience profound dissociative states, reporting that hours and even days have passed by like minutes. During periods of extreme dissociation they may perform a variety of adaptive and maladaptive behaviors and may harm themselves or attempt suicide.

The idea that many people have a hidden "childlike" or regressed personality which emerges and takes control in specific circumstances has been prevalent in the popular media. The many adult and children's cartoons and movies that depict a mischievous little "being" inside a person that causes them to do things against their will and with whom one argues is a good example. The "little one" is usually depicted as the manifestation of the person's unconscious forbidden urges and desires, or the devil (i.e., "The devil made me do it!"). The part of the personality that argues with the little being is represented as the conscience (the responsible adult).

Early and modern hypnosis researchers accepted the idea that some hypnotized subjects exhibited a divided or co-consciousness or a doubling of the self (Binet, 1890; Dessoir, 1890; Prince, 1909; MacMillian, 1977; Hilgard, 1979). Researchers discovered that some subjects experiencing conversion anesthesia could not report a cutaneous sensation, but, when whispered to, could accurately describe the location of the stimulus (Hilgard, 1979). William James (1889) reported that a subject whose arm was anesthetized under hypnosis accurately reported via automatic writing the number of times the arm was pricked after being asked to do so in a whisper.

Hilgard, et al., elaborated on this theme and demonstrated that some hypnotized subjects possessed at least two personality parts. The 'hidden observer' part allowed the hypnotized part to carry out the suggested behaviors but also observed what was happening. According to Hilgard, the hidden observer is the normal observing part that is in control in a normal non-hypnotized state. It is objective and reality oriented.

Hilgard demonstrated that some hypnotized subjects, while oblivious to pain, were still processing information such as the relative intensity of the painful stimulus. The 'hidden observer' was monitoring the increments in the painful stimulus while the hypnotized or dissociated part was not experiencing the stimulus. There are many examples of this phenomena occurring during hypnosis; e.g., when somnambulistic subjects are given the suggestion that when they open their eyes all the furniture in the room will have disappeared (negative hallucination), they express surprise to see people sitting on air. However as they walk around the room they avoid walking into the furniture although still reporting that they do not see the furniture. On one level they are aware of the furniture while yet denying its existence on another. This doubling of consciousness probably explains why hypnotized subjects will not commit an act that is contrary to their conscience even while in the deepest hypnotic state.

The following is a subject's description of her personality structure while participating in Hilgard's, et al., demonstration of the 'hidden observer' phenomena:

> "The hidden observer is a portion of Me. There's Me 1, Me 2, and M 3. Me 1 is hypnotized; Me 2 is hypnotized and observing; and Me 3 is when I'm awake. The hidden observer, a part of Me 3, is cognizant of everything that's going on; He's like a guardian angel that guards you from doing anything that will mess you up... Unless someone tells me to get in touch with the hidden observer, I'm not in contact. It's just there. " (p. 71)

Throughout the history of psychology, MHPs have described clients who regress to childlike behavior when stressed. Most lay observers would agree that many intelligent and knowledgeable people act like a child when stressed. Many people spontaneously regress to an early mode of conflict resolution when stressed. Sometimes clients spontaneously regress and speak in a childlike voice during abreactive extinction. The extreme stubbornness and reliance on trance-logic that one witnesses in most chronic A-Bs, OCDs, and ADs is also typical of the thinking style of an intelligent, imaginative, but emotionally upset child. Hysterical personalities often function in an obviously childlike manner and may derive much secondary gain from this behavior. As stated, women in most societies are rewarded for behaving like little girls. The stronger the hysterical features, the more regressed the client behaves. The observation that A-Bs often enter trance-states and feel like someone else is in control supports the hypothesis that these states are similar to hypnotically regressed states.

Dissociative states and hypnotic states are so similar that they may be qualitatively the same. Hilgard's, et al., neodissociation theory of hypnosis supports my view. He proposes that dissociation varies from limited and mild to wide-ranging and severe. People also vary greatly in their ability to dissociate. Some are able to produce simple superficial responses only to direct suggestion, and others are able to produce massive dissociations that appear to be pathological conditions such as fugue states and Dissociative Identity Disorder. What determines a dissociative disorder is when the dissociations become widespread and significantly hinder a person's functioning.

Trance therapy, originated by J. Hevesi and described by Zeeman (1978), is essentially free association in a light hypnoidal state. Trances are utilized as a means to integrate the A-Bs' fragmented personality. During trance therapy, A-Bs relax in a dimly lit room with the therapist for an extended period of time. Through free association or spontaneously, an immature or regressed part of the personality emerges into the clients' imagination. Clients converse with the fragmented part and together the parts formulate a plan to work their way out of the problem. Hevesi reports that many A-Bs state that the 'monster' inside of them takes over and devours food for several hours (Zeeman, et al.). Zeeman's descriptions are in agreement with my observations that A-Bs often regress to a fixated or dissociated personality during the trance state which takes control and initiates the binge-eating and vomiting or starvation.

Zeeman, in his description of Hevesi's work, explains the etiology and treatment for A-B in terms of the mathematical catastrophe theory. According to Zeeman, in order to alter the maladaptive behavior the MHP has to create a third behavioral mode or state during which the basic insecurities of the client can be treated with reassurance. The new behavioral mode must lie between the abnormal extreme states of mind that correspond to the A-B states. Zeeman states that the catastrophe theory predicts that A-Bs spontaneously fall into fragile trance-like states. Hevesi's treatment consists of about twenty sessions, each lasting one to three hours, during which clients enter trance states and he helps them reconstruct a normal state.

Hevesi and Zeeman recognize that hypersuggestibility characterizes these dissociative (hypnotic) states that A-Bs spontaneously enter. The transition state between waking and sleep corresponds to what Budzynski (1976) termed 'the twilight state.' Evidence exists that during this twilight state (characterized by the theta EEG pattern) learning occurs rapidly (Budzynski, et al.). This may explain why casual comments by the MHPs are influential, but this is purely speculative. They begin their therapy by emphasizing that having or trying to get the A-Bs to eat is not useful. They feel that this frees A-Bs from being occupied with their fears, and allows them to relax and look at themselves objectively. Clients are then told to look in a non-emotional way at what their minds produce. This reminds me of Buddhist mindfulness meditation that has been incorporated in behavioral therapy by Romer and Orsillo (2009).

Most authors (Zeeman, et al.; Crisp, 1980) agree that A-Bs use their maladaptive behavior to withdraw from and blunt a variety of negative emotions. As emphasized, the temporary inhibition of the negative mediating emotions by slipping into dissociative states guards these emotions from extinction, and ensures their recurrence. Operant conditioning easily explains the development of the automatically occurring trance behavior. The act of slipping into dissociative states is an operant response and is reinforced by the temporary reduction of the negative emotions that follow the trance behavior.

Zeeman states:

"The patients confirm that when they awake from the first few trance sessions they find themselves sometimes in a fasting and sometimes in a gorging frame of mind" (p. 57).

I have witnessed this phenomenon following hypnotic trance-inductions as the abreactive extinction procedure involves the purposeful elicitation of these feelings. Zeeman describes that A-Bs experience themselves as 'double personalities' - one personality called the 'real self' and the other is called various names such as 'the little one,' 'the imp,' or the 'demon.' Hevesi stimulated communication between the two parts. For example, the 'real self' promises to take care of the other part if the starving is stopped. After repeated trance sessions, a fusion of the parts occurs and the clients emerge into a more normal state, and the starving is stopped.

Zeeman's descriptions are congruent with my conclusion that A-Bs employ dissociation as their major defense mechanism and often manifest fragmented personalities. However, I have found that the severity of the A-B behavior may not correlate to how well the client succeeds in therapy.

Zeeman agrees with me concerning the importance of what he terms basic insecurities and what I term negative mediating emotions (CERS), which I contend are derived from separation anxiety or trauma. In addition to integrating the personality through trances, I stress the systematic extinction of these emotions. Zeeman makes little reference to anything similar to the extinction process.

Separation therapy (Kirsten & Robertiello, 1975) and ego-state therapy (Watkins, 1979) employ concepts similar to trance therapy. In separation therapy, clients are instructed to use their imagination to divide the personality into three distinct parts, "little," "big," and the "individual." "Big" (the adult part) and "little" (the child) carry on a dialogue while the "individual" acts as a referee. The adult satisfies the child's wants and needs, but also sets reasonable limits on the child so she does not get out of hand. In other words, the adult part treats the child as a loving intelligent parent would. Successful case studies have been reported for a variety of disorders using this method.

During separation therapy, the fragmented part of the personality is experienced as separate from the self, and with the help of the therapist, the client and her "adult" part make the fixated fragment secure. In my opinion the "child" part is the fixated or regressed representation of the overly learned motivations and expectations that represent the resistances encountered when treating chronic clients. The age of the fixations may correspond to traumatic experiences involving specific traumas such as sexual molestation, but the strong emotions that caused the fixation and maintain the fragmentation usually involve separation anxiety. However, whether the mediating CERs were due to traumas, separation anxiety, etc., the result is the same and the treatment is the same.

Kirsten's, et al., technique is useful in the treatment of A-B and lends itself well to hypnotic procedures and trance states. The hypnobehavioral techniques I use to integrate the personality fragments are similar to Hevesi's techniques, and this integration is facilitated by the abreactive and in-vivo extinction of the emotions that caused the fragmentation. With extremely resistant clients, abreactive extinction by hypnotic regression may be the best and quickest way to extinguish the negative emotions that cause the client to resort to dissociation. Following the abreactive extinction of the mediating CERs, most clients find they are experiencing a significant decrease of the urges to starve or binge-eat and vomit by the third day of treatment. By preceding FRP with abreactive extinction, the flooding is made more tolerable since part of the negative CERs have been extinguished.

I have encountered a few clients who have resisted hypnosis, trance therapy, and separation therapy but still experienced complete remission of their long-standing maladaptive behavior as a result of only in-vivo FRP. One such client, who was in her early 50s, underwent FRP for only two days. At a three year follow-up, she was almost entirely free of her binge-eating and vomiting and did not feel the need for further treatment.

Most chronic A-Bs are profound somnambulists and, in some, the hypersuggestibility was induced by traumatic early experiences. The few A-Bs I have encountered who had experienced a traumatic sexual molestation early in life usually exhibited extreme dissociation as a coping style and took longer to successfully treat. Early sexual traumas seem to increase a client's use of dissociation as a defense, and they develop more somnambulistic behavior. However, many chronic clients are encountered who are somnambulistic and have not been traumatized. I have no explanation for this other than that they inherited a strong hereditary predisposition for dissociative and OCD behavior. These clients are often the most resistant because they rely heavily on trance logic and trances as coping behavior.

Those who also exhibit psychopathic characteristics in that they are extremely self-centered and do not understand the concepts of love and guilt are more difficult to treat. These clients would best be described as possessing a rigid inherited character disorder which will most likely not respond to even the best designed treatment approach. I encountered two women diagnosed as borderline personalities and had no success with either client. Anxiety and its physiological components are controlled by midbrain structures, whereas characterological disorders have more to do with the frontal lobes.

The data concerning the role of heredity in the origin of A-B, AD, and OCD behaviors show some hereditary predisposition to these disorders (Butcher, Minder, & Hooley, 2010). Somnambulism (hypersuggestibility and trance logic) may either be an inherited trait or an acquired ability. Early traumatic experiences or an unloving environment most likely force an imaginative child to develop hypersuggestibility and somnambulistic-dissociative behaviors as a means of coping, but this conclusion is speculative.

In summary, traumatic sexual molestations and incestuous experiences are extremely difficult for children to accept and cause them to resort to dissociation. Natural somnambulists inherently resort to extreme dissociation in order to cope with negative feelings without having experienced an extreme emotional trauma and are, therefore, often difficult to treat. Somnambulists are more fully described in the following chapter.

L. Hypnobehavorial Models of Obsessive-Compulsive Behavior and Unconscious Motivation

A-Bs' distorted body image serves their unconscious and conscious desire to starve. Through hypnosis the fragmented part of the personality that controls the starving or binge-eating and vomiting may admit that she is purposely causing the conversion reactions in order to keep the adult from eating. During hypnosis the client is instructed to question the fragmented part aloud and, without analyzing, listen to the answers. Clients are often surprised at the answers. The dissociated part may be using the A-B behavior as a way of punishing the adult part or a significant other for rejecting her or not attending to her needs for love and security. Occasionally the fragmented part is trying to murder the adult part. Some clients are so fragmented that they would be classified as having a multiple personality dissociation identity disorder. By repeatedly triggering the various personality states, the client can begin to integrate her personality. The various fragments become aware of each other and then integrate into one self. By the client understanding the motives and goals of the other part(s), integration and therefore better self control is facilitated. Abreactive extinction also strongly facilitates integration because the CERs that mediate the fragmentation are greatly decreased.

Again I must caution MHPs concerning hypnotic regressions and all techniques that allow clients to slip spontaneously into trances and encounter past traumas. The recall of the traumas varies from being highly accurate to total fantasy. In the vast majority, it is impossible to know. Refer to Section I for more details.

In my approach, the regression to past traumas is used only to form an affect bridge between what the client is experiencing now and when the trauma began. This is done to abreact and extinguish these negative emotions. When these are encountered, I am not interested in validating whether memories of traumas are accurate or not. Human memory is plastic and unreliable, and accuracy of memory is not important for extinction to occur. The goal is extinction, not insight!

I thought MHPs would have learned their lesson concerning the validity of recovered memories that are supposedly repressed. Some MHPs believe that 80% of eating disordered clients have been molested (Schwartz, 1998). They emphasize insight instead of extinction. Their continued probing to uncover repressed traumas suggests that molestation occurred to already hypersuggestible clients. This insistent probing results in prolonged therapy which is harmful to the client and produces a secure stream of income for the MHP.

In December 2011 MHPs were being sued for implanting harmful memories in A-Bs at the prestigious Castlewood Treatment Center. The Co-Director, Mark Schwartz, Ph.D., stated in a 1998 interview published in St. Louis Magazine that he and his wife treated 3,000 clients and more than 80% had been sexually abused. Of course there is no way to validate that he and others were not inadvertently helping these hypersuggestible people to recover false memories. (Cara, 2012)

No MHP gains financially by getting clients better quickly. Instead they are rewarded by keeping clients coming for as long as possible. When I had an office at one medical center, a colleague remarked, "You can make fun of the Freudians, but who is really stupid? You cure these people in six days and the Freudians keep them coming five times per week for years. They have a steady secure income, while you have to keep relying on referrals." My colleague was proven right! Referrals stopped and luckily I had other sources of income.

The reality of this should be obvious. One client who is suing Castlewood spent 15 months in residential treatment and was billed $650,000. Since this client is a resident of Minnesota, insurers by state law must pay for long-term treatment of eating disorders. This client's suit claims she was led to believe she had been involved in satanic cult activities during which she was raped multiple times, forced to eat babies, and had twenty different personalities (Cara, 2012). Interestingly, the FBI and other law enforcement organizations cannot find evidence to support the above mentioned crimes.

As stated, the fragmented personality can usually be integrated by extinguishing the emotions that are maintaining the fragmentation. Once these powerful feelings are extinguished, the personality is freer to integrate and may spontaneously do so. However, the negative emotions and associated memories that caused and maintain the fragmentation may be overpowering and cause clients to dissociate them and deny their existence. Partially successful denial guards these emotions from extinction, allowing them to surface when the defenses are weak. The following case is an example of how the negative emotions acquired during one traumatic incident mediated maladaptive behavior not directly related to the incident.

This client entered therapy because of repeated uncontrolled binge-eating and extremely low self esteem. Although she had a past history of A-B and had had to be hospitalized during adolescence, she had never vomited after binge-eating. The client was in her late twenties, employed in a responsible position, and had maintained an independent adult existence since her graduation from college. Her description of her parents indicated that, although somewhat rigid, they were usually warm and loving. In addition, she was sexually inhibited, although interested in forming a close emotional and sexual relationship with a man. Because she was very religious, her sexual inhibitions appeared to be the result of religious guilt. Her binge-eating usually occurred when she felt alone. The antecedent learning situations became evident during her second three-day therapy session.

When she was four years old she had crossed a busy street without her mother's permission to look for a friend. Instead of finding her friend, she encountered the friend's father, who persuaded his 13-year-old son to rape her while he watched. The raping was, of course, anxiety eliciting and caused her to become hypersuggestible. During the raping the father and son both told her that she was a 'bad girl,' warning her that if she told her mother about what had happened, her mother would not love her. Her mother had been looking for her, and found her just as she left the friend's house immediately after the raping. Her mother was understandably upset because she had crossed the street alone, and had spanked her while calling her a 'bad girl.' The client never told her mother what had happened because she had accepted what the man and his son had said, and her mother's actions were interpreted as a confirmation that her mother would not love her and that she really was a 'bad girl.' This instance likely accounted for a large part of her negative self-concept.

The first three days of therapy involved FRP, hypnotic regression, abreactive extinction, and imagery conditioning, which resulted in a complete remission of her symptoms for approximately six months. The strong anxiety that resulted from the molestation remained dissociated until she encountered new stressful situations. These stressors, which were unrelated to sex or men, caused her to again experience overwhelming panic attacks and urges to binge-eat.

She returned for an additional day of therapy during which the sexual molestation was recalled through hypnotic regression. She realized that the emotions she felt during the abreaction of the molestation were the same emotions that preceded her binge-eating. Through repeated abreactive extinction, the CERs mediating and reinforcing the binge-eating were extinguished.

In this case the major etiologic variable was not a faulty family structure, but the sexual molestation and what the client had interpreted as rejection from her mother. It is easy to see why she viewed herself, even as an adult, as a 'bad girl' and still feared her mother's rejection. As a result of this belief, she had become compulsive and perfectionistic. No matter how much she achieved, she never felt good about herself.

The bulimia decreased after her second visit, and she was able to lose her excess weight. The client later discussed with her mother what had happened to her as a four year old; unfortunately the mother refused to believe that it had happened and told her daughter that she most likely only imagined it. After encountering her mother's disbelief, she again began to deny the sexual molestation and her bulimic behavior immediately returned. The converse also occurred: whenever she accepted the reality of the molestation, her urges to binge and resultant bulimic behavior immediately subsided. Over the following year, the client reported that her baseline anxiety had greatly decreased and had only experienced one setback--- her binge-eating had again returned due to stress experienced during changing jobs while simultaneously again denying the molestation had happened. (Her mother later revealed that she had heard from reliable sources that the man in question had impregnated his daughter.)

Hypnotic regressions revealed that food had been associated throughout her childhood with love and was used as a reward for being a 'good girl.' She therefore sought food as a tension reducer whenever negative feelings emerged. In many cultures food has acquired relaxing properties and come to symbolize parental love. Eating is of course also inherently enjoyable, and food is easy to obtain in westernized society. It is also more socially acceptable and less dangerous immediately for someone to reduce tension by eating or compulsively working than by drinking or taking drugs. After a total of six days of therapy, she rated herself on a two-year follow-up as 90% improved.

Self-destructive unconscious motivations were easily uncovered in a 32-year-old anorexic who had been living on the brink of starvation since adolescence. She had been referred for therapy by a local physician because of headaches and her starvation. The client had been divorced twice, lived with her teenage daughter, and was employed as an LPN. She experienced minor fugue states and rapid mood changes.

Through hypnotic regression, it was uncovered that when she was 9 years old she had been angry with her sister and wished that her sister would die. Unfortunately the sister was killed by a truck the following day. The trauma caused her to fragment her personality and to blame and punish herself. It is easy to understand how a suggestible and imaginative 9-year-old child might believe that what she imagines influences physical reality. The fragmented regressed part was contacted in the first two hours of therapy during the hypnotic regression.

The following is the dialogue between the client and myself following the history taking, explanation of hypnosis, and a short hypnotic induction:

DK: I am going to explain some of the basic laws and rules of how your mind works so that you may apply these laws and rules to yourself. Remember that knowledge is power. You're going to begin to understand yourself and understand your own subconscious motivations. Most of all, you are going to begin to accept yourself and love yourself. Too often we judge ourselves for being human -- for having human emotions, feeling things, thinking things -- and it's not our right to judge ourselves and to punish ourselves. This is not our right. Remember, we are all human.

So as you're relaxing now, the first law I'm going to explain concerns behavior: all of our behavior follows our expectations. As we expect to become, we become; as we expect to behave, we behave. These expectations and images of ourselves which is our self-concept and which we automatically follow, whether it has to do with eating, what we do in life, whatever, is implanted into our minds through our imaginations, not our willpower.

You know how weak the willpower is compared to the imagination. Think about the problem you've had concerning eating... think about it: no matter how hard you try to block or push away those feelings of nausea, those bad feelings, it doesn't work.

Nod your head if you agree.

(Client nods.)

Fine. That's because it is in your unconscious mind, and these feelings of nausea were put in your mind through your imagination, through suggestions. For example, when you try very hard to relax, what happens? You become tenser. So forget about trying. You have a good mind and a good imagination. You're going to learn to use that imagination and your mind to help yourself.

The client was then regressed and told to go back to the "feelings, thoughts or time when the problem began," or to the part of the personality that was underlying her eating problems. The fragmented part from that age was easily encountered, and the following dialogue took place:

DK: So, you're 9 years old and you say that she (the adult self) doesn't deserve to live? Tell me, why she doesn't deserve to live?

C: She killed her sister.

DK: How did she kill her sister?

C: She wished she was dead, and the next day she was.

DK: How old were you when this happened?

C: 9

DK: So... you accepted this when you were 9 years old. Now just look... you're there, right now; you're 9 years old... What killed your sister?

C: A car... no, a truck, I think.

DK: You thought that by wishing your sister was dead that you killed her, didn't you?

C: I did!

DK: Did you?

C: I wished it!

DK: I know you're only 9 years old and this is how a 9 year old thinks, but let me tell you something: you didn't kill your sister. Your wishes are not that powerful. But you thought you did, because it happened very soon after you wished it. It's very normal for a 9 year old to be angry at her sister, and if your sister had lived, you wouldn't have had this problem. What would have happened is that you would have ended up loving your sister and you would be good friends today. But instead, what happened did so just by chance -- by chance your sister died very close to the time you wished she were dead. So now you're killing yourself because you feel responsible, aren't you?

C: Yes.

DK: Do you believe in God, little girl?

C: Sometimes.

DK: Well, you know what God has told us? That it is not our right to judge or punish ourselves or anyone else. So, you don't have the right to judge or punish yourself. Let that be God's job. He's not going to punish you for your wish -- every child has wishes like that.
(Pause)
What are you thinking about now?

C: My father.

DK: What about your father?

C: I don't like him.

DK: Why don't you like him?

C: I saw my daddy with my mother when I went to their room.

DK: What were they doing?

C: Having sex.

DK: Okay... and you assumed what? That it was bad, that your father was hurting your mother?

C: Yes.

DK: You assumed he was hurting her? Tell me why...

C: She was aggravated, I could tell by the look on her face. She didn't want to have sex.

DK: Do you still dislike your father?

C: Yes.

DK: Why? You don't have to remember any of this when you awaken, that's up to you. Just tell me why you dislike your father.

C: The way he treats his family.

DK: How does this relate to your stopping eating?

C: It doesn't...

At this point the client began to come out of the trance as more emotionally charged material began to surface. She had been raped at 17 and, although the raping had not been denied, the associated emotions had never extinguished. These emotions were abreacted and extinguished before continuing with the problems that had occurred when the client was 9 years old.

DK: Now, go back to being 9 years old... Nod your head when the 9 year old is there.

(Client nods)

Okay... now tell me... are you punishing the adult by starving her?

C: Yes.

DK: Are you going to continue to do it now?

C: Yes.

DK: Why?

C: I don't like her.

DK: You don't like her because you think she killed your sister, correct?

C: ... I don't like her.

DK: Why?

C: She's not anything.

DK: You blame her for killing your sister, don't you?

C: Yes.

DK: And that's why you started starving her, right?

C: Yes.

DK: And now you are still starving her. Why? -- For the same reason, you think she killed your sister? That she wished her sister was dead? Is that part of the reason you continue to starve her?

C: Yes.

DK: You're the one who comes in and takes over when she ends up someplace else and doesn't know how she got there, when she has those lapses of memory, aren't you?

C: Yes.

DK: And you're going to punish her for that also?

C: Part of it.

DK: What's the other part?

C: She's a failure.

DK: So... you're going to make sure she's a failure, aren't you?

C: Yes.

DK: So you make her a failure and then you punish her for being a failure. Do you understand what you are doing?

C: She is a failure, anyway; she's scared of people.

DK: Who makes her scared of people?

C: She does.

DK: No, you do. Who brought on the nausea so she couldn't eat? Who did that?

C: I did.

DK: That's right, you did. I'm not mad at you about it. You're a very nice 9-year-old girl, but you have some ideas that are wrong. You didn't kill your sister; you had nothing to do with it at all. All children wish at sometime or another that their sister or brother would die when they're angry with them. Do you realize that's normal?

C: No.

DK: Why do you say no?

C: Because I loved her.

DK: That's all right; we can love and hate people at the same time when we get angry. Were you angry with your sister at that time?

C: Yes.

DK: And you wished she was dead then, right? And just by chance, she happened to die -- which was very tragic -- but, you didn't do it. But you're going to continue to starve her? Is that what you're telling me?

C: Yes.

DK: You need to start to understand something, and then I'm going to ask you to make a deal with me. You see, you are her -- do you know that you are really her? And, she's you. Do you know that?
(Pause.)
Tell me... talk to me...

C: No, I'm not.

DK: You're not what?

C: I'm not her.

DK: Yes, you are. Now, you do want to keep your secret, right?

(Client nods.)

You can keep your secret -- I won't expose you to the world, because you don't want that, do you?

C: No.

DK: And I have no desire to do that because you're 9. There's no reason to punish you or anything like that, because you didn't do anything wrong. You know your feelings are hurt, and you thought you did something you actually didn't do, but you're going to have to make an agreement with me right now. You have to agree to begin to eat a little bit more of the good foods, just a little bit more, all right? Then I won't have to expose you.

Now, is that a deal? You understand why I have to ask this, don't you? I can't allow you to kill yourself because that's morally wrong. Your sister died, and if you also died, it would be just another bad event, wouldn't it?-- two lives lost. Do you understand what I'm saying?

C: Yes.

DK: So understand that I don't want that to happen. If one person dies, and then we lose another person, it becomes another tragedy, understand?

C: Yes.

DK: So I can't sit back and watch you do this to yourself. I won't tell the world on you. I have no desire to hurt you; I want to help you. So I won't tell on you as long as you agree to begin to eat a little bit more, all right? Now... will you promise me this, little girl?

C: (Nods her head in agreement.)

DK: Good, because I like you and I don't want to punish you or anything like that, because you didn't do anything wrong. So I am not here to punish you, I'm just concerned about life, because you're the one who is dying. I know you don't believe that right now, but, you are dying too. Do you want to die?

C: No.

DK: Good, but you're going to have to change your belief that you caused your sister to die, because making yourself die goes against everything that's right, do you follow me? You may not understand now, but you will.

Again... you're going to agree now to begin eating a little bit more. Are you going to promise me this?

C: Yes.

DK: Very good. Now, when you awaken, do you want to not remember this conversation? It's up to you.

C: No, I don't (want to remember).

DK: You don't want to remember it? All right, I'm going to awaken you, and you won't have to remember our conversation at all, okay? You're going to feel good and you're now going to start eating a little bit better, right?

C: Yes.

DK: Okay, that is our agreement then.

Now... relax... going deeper into relaxation, deeper on down... Now, when I awaken you, you're going to feel wide awake, feeling relaxed and feeling good, not bothering to remember any of our conversation. You're going to feel good, you're going to be relaxed, and awake.

The client was then awakened.

DK: How do you feel?

C: (somewhat sleepily) All right...

DK: Look at me now... You're wide awake. You'll find it a little bit easier to eat now, okay?

C: Okay...

DK: You're going to begin to eat just a little bit more so you should begin to maintain your weight, for health and for life, all right? Do you remember what went on?

(Client shakes her head no)

That's okay because you're going to eat a little bit better now, and each time you go into hypnosis with me, you're going to find it easy, like you did today.

C: Did I cry?

DK: Yes, you cried a little. Now, I want you to see me once a week for a while. If there is any problem in between appointments, I want you to call me. You aren't going to listen to the tape I made today; you and I will listen to it sometime in the future. You will be using some hypnosis at home in the future, but I'm giving you this next week to relax. You'll begin to eat a lot better, and you'll be starting to feel a lot better too.

C: But I don't want to relax, I want to get this show on the road!

DK: The show is already on the road.

C: It is?

DK: Yes, it's on the road now, because you and I have an understanding. Relax now, and you'll find you will be eating a little better over the next week.

During the following week the client experienced four days of normal eating and gained three pounds, from 92 to 95 pounds (she was 5'5").

This case illustrates how a fixated part of the personality may function independently of the adult and be bargained with for a temporary cessation of a maladaptive behavior. Of course at this point she was far from being 'cured.' More sessions were needed in order to extinguish her anxiety, anger at men, and the unconscious motive to kill herself.

This client was seen four times over six weeks. Following the fourth session, she voluntarily stopped treatment because she felt that it was 'not working fast enough' and was too expensive. Two months after termination, I received a letter from a plastic surgeon asking if this client would be adversely affected by breast augmentation. Her desire for this operation might lead one to believe that she was more motivated for cosmetic improvements than to be rid of her maladaptive behavior. Clients who are unmotivated or expect the therapy to work without putting forth effort are of course poor candidates for treatment.

This particular client did return three months later for an additional two sessions. In the interim, she had consistently been using the hypnotic tapes that I had made for her and realized the suggestions concerning self-love, and relaxation were helpful, as her eating behavior had improved considerably. Her weight was stabilized at 100 lbs., she was now interested in and enjoyed dating, and overall she was feeling better about herself. Two years later she reported that her weight was still 100 lbs., she was maintaining a good diet, and was no longer depressed.

Returning to the topic of conversion reactions, they are predictably elicitable and follow the parametric characteristics of respondent conditioning and extinction. A-Bs experience the psychophysiologic autonomic and somatic responses that are concomitant with classically conditioned fears, including vasoconstriction throughout the arterial tree, increased palmar sweating, increased levels of tension throughout the trapezius and frontalis muscles, and the subjective feelings of a knot in the stomach and nausea. These responses are also extinguishable through abreactive and in-vivo extinction. Again, biofeedback training is also very helpful in alleviating these responses.

The conversion reactions are often elicitable by simply having A-Bs put a piece of food they consider fattening into their mouth, by just touching it to their lips, or by having them imagine they are eating the food. By having them imagine the eliciting stimulus situations or by encountering the eliciting stimuli in-vivo, the conversion reactions and fears can be repeatedly elicited and extinguished.

There may be many stimuli that elicit these reactions but they usually involve food, feeling rejected, being alone, boredom, the thought of becoming fat, or even gaining a fraction of a pound. Other stimuli such as any negative emotional experience, the thought or sight of fattening food or fat people, tight clothing, or the bathroom scale may also elicit these reactions. As with all conditioned responses, these reactions tend to generalize to similar stimulus situations and become elicitable by an ever-widening range of stimuli.

It is enlightening to A-Bs to discover that simple act of imagining the relevant stimuli can cause the conversion reactions. This indicates that it is not the food itself but their imagination that is causing these reactions. I explain to clients that since suggestion and imagination can bring these feelings on, suggestion and imagination can take them away. This enables clients to understand that they control and change their emotions by using their imaginations and hypnotic skills. Experiencing this self-control is beneficial because it instills an attitude that is incompatible with the fear of losing control, which is common to A-Bs and OCDs.

The conversion reactions of feeling fat in the stomach, thighs, etc., are distorted sensory experiences that are the result of the expectation that food will make them fat and that the fat is immediately deposited throughout the body. It is easy to see how trance logic (the ability to accept logical incongruities and be unaware of them) (Orne, 1959) is employed by these clients. They simply accept that food means calories, calories mean weight gain, and weight gain means rejection, which then elicits separation anxiety. The contiguous pairing of these thoughts and emotions cause them to automatically elicit each other.

The distorted body image of being fat, irrespective of the actual body weight, in combination with the unrealistic expectation that even the smallest amount of food will make them fat and their uncanny ability to dissociate emotions while accepting logical incongruities are the important variables maintaining A-Bs maladaptive behavior. The following cases are examples of how intelligent educated women used trance logic to govern their behavior.

A registered nurse once argued that the bloating sensations and feelings of being fat that she experienced immediately after touching food to her lips were the result of water retention. She insisted that she had gained a considerable amount of weight overnight, although she had only drunk a few glasses of water and eaten nothing during the prior 24-hour period. She refused to weigh herself, however, because she feared gaining weight.

Questioning revealed that the 'fat feelings' had been elicited by her feeling rejected from her boyfriend. Reviewing with her the the history of the specific incidents that led to the elicitation of the conversion reactions showed her the cause of her feelings. She admitted that if her patients acted the same way, she would have told them they needed to understand that pounds of weight could not be gained overnight if intake of food and liquid had been severely restricted. However, when it concerned herself, she was more willing to trust her feelings (conversion reactions) than any of the material she had learned from her courses.

She was unable to, or refused to relate her book knowledge to herself, even after having experienced three days of therapy during which she had been able to elicit these same feelings through imagery and hypnotic suggestion. Instead of facing the strong negative emotions elicited by what she perceived as rejection from her date, she relied on the trance logic explanation that, "I was rejected because I must have gained weight; therefore I must starve in order to lose the weight so I will be accepted." She returned for a second three-day therapy session and has since been reasonably free of the starvation behavior for over two years.

Another A-B who had a Ph.D. and a responsible administrative position in a large university believe that if she ate high protein foods first and then junk foods, the protein would have more time to digest as it would be in the 'bottom' of her stomach. She, therefore, reasoned that when she vomited, mainly the junk food would be eliminated. Questioning proved that she was aware that the stomach churns food and that the complex proteins and carbohydrates are digested more slowly than the refined carbohydrates in the junk food, but she could not relate this information to herself. She had governed her behavior by accepting this incongruity for over seven years. This insight and the extinction procedures only temporarily caused a decrease in her bulimic behavior, and she experienced significant relapses after returning home.

About one year later she reported that she needed to come in for her second three-day therapy session. I reviewed the concept of extinction over the phone with her and found that she either had not understood the concept or was unmotivated to understand it at the time of the initial three-day session. She appeared to understand it during the phone conversation and agreed to practice the extinction in-vivo before returning. She experienced significant improvement on her own and therefore did not have to return for the second three-day therapy session. This client likely had understood the concept initially, but due to a variety of conscious and unconscious reasons, she had dissociated that knowledge. Her motives for keeping the behavior apparently changed by the time of the phone conversation, and she was finally ready to use extinction procedures to her advantage.

As just illustrated, clients may thoroughly understand on a cognitive level the theoretical explanation of conversion reactions, but refuse to apply this information to their own lives. If they do not want to believe something, they will simply ignore it and convince themselves that it does not apply to them. As with most OCDs and ADs, questioning A-Bs' illogical beliefs leads to agitation and more defensive behavior. Some of this defensiveness can be avoided by thoroughly explaining the concept of trance logic to them. The explanation is usually better accepted if the therapist also explains that many creative writers and artists use trance logic in order to be creative, but that this ability must not be allowed to get out of control as it can make them a victim of their own imagination. Explaining trance logic and its devastating consequences can also be helpful in illustrating the hypnotic law that "All behavior follows our expectations. It is the imagination, not the willpower, that implants those expectations into the unconscious mind." Examples of this law should be given.

Many people, particularly somnambulists and children, may accept an obviously illogical inference and then govern their lives by it. As people become anxious, they become more intolerant of information that is dissonant with their beliefs and rely more on faith and trance logic. Heightened anxiety tends to make them less willing to abandon their illogical and maladaptive beliefs that in the past were associated with security. This same phenomena has occurred in whole societies. When people become confused and insecure, they often look for simple explanations or an authoritarian leader to guide them. Rigorous critical thinking is impossible for some, even those with high IQs.

It is bewildering, frightening, and fascinating to view educated intelligent people compartmentalizing knowledge, dissociating that knowledge from the rest of their knowledge, and then acting as if that knowledge does not apply to their reality but only to the reality of others. The inability or reluctance to view things more objectively and honestly is the reason for many ridiculous applications of unsubstantiated methods and ideas in our educational system and methods used by MHPs. This phenomenon has been viewed in college professors who expound and act upon ridiculously unrealistic interpretations of how the world functions.

I have encountered many racial, political and sexist bigots, regardless of sex, race or IQ. In college environments one-third to over one-half of college professors admitted that they would discriminate against a conservative colleague, even though they intellectually claim stereotyping is unreasonable and unsupportable. Among psychologists, this bias is even higher. How many others do discriminate and will not admit to it or are unaware of their bias? These professors have no appreciation of equal treatment before the law, but guide and rationalize their behavior by trance logic to suit their own feelings. The very people who are supposed to be free of bigotry are often the biggest bigots (Inbar & Lammers, 2012).

Once somnambulistic subjects accept suggestions, whether logical or not, they will hold on to them with extreme tenacity. The more anxious they become, the more dissociation they employ and the more resistant they become to accepting dissonant information. One would think that during dissociated states the client would become hypersuggestible and therefore more open to positive suggestions. Unfortunately this often is not the case. In fact the somnambulists' ability to dissociate and enter hypnotic states is often correlated with the stubbornness of the maladaptive behavior. As stated previously, once they accept an idea it is very hard to alter that idea. Once the somnambulists are attached emotionally to other people, it is also very hard to alter that relationship.

Although I disagree with the Spiegels', et al., abandonment of hypnosis when working with somnambulists, my experience is in accordance with their view that these clients are difficult to work with because they use their trance ability against themselves, therefore limiting positive change.

Highly hypnotizable individuals, because of their extreme trance ability, often present a frustrating challenge to MHRs.

Highly hypnotizable clients in their normal state exhibit stronger than normal tendencies to trust others and to rely on trance logic. They have great capacity for intensely focused concentration to the degree that they can become so absorbed by what they are doing that hours go by like minutes. This experience of time distortion is also characteristic of the state of dissociation that bulimics experience when they are binging and vomiting. Highly suggestible clients easily produce on their own all the phenomena experienced in hetero-hypnosis. They can easily regress to spectacular and what seems like important episodes in their early life, but do not connect them with their present experiences.

They appear to possess a wonderful imagination, and one would think that they would use this ability to lead great creative lives. However, they are often a victim of their own trance-imaginative abilities and the ability to be unaware of logical incongruities. Lastly, highly suggestible clients when stressed can develop what appears to be a psychotic state. During this state they can hallucinate and appear almost schizophrenic. However, this is a pseudo-psychotic state and can be manipulated with hypnotic suggestions. Often MHPs confuse this with schizophrenia.

MHPs who have encountered resistant A-Bs and are aware of the difficulty in obtaining their trust may be confused by somnambulists' proneness to trust. This paradox is understood by realizing that once somnambulists accept an idea (i.e., whether to trust or love a particular person or not), they will hold on to that idea with great tenacity. As stated, somnambulists also tend to trust their conclusions whether logical or not, and trust their emotions. This explains how they can become disastrously enmeshed in pathological relationships.

Somnambulists can easily regress to spectacular and seemingly important past experiences that are isolated and not relevant to their current lives. They easily produce conversion reactions and dissociate episodes during which ABs can eat but not remember their actions later.

Just like the nurse who had convinced herself that if she ate proteins first they would be vomited last, these clients can accept an illogical autosuggestion and guide their lives by them. They are often unconscious of the extreme logical incongruities of their maladaptive beliefs. One would think that clients who can produce such intense experiences would be very flexible and possess remarkable abilities to change. However, just the opposite is often true. Underneath this flexibility and creativity is often a fixed core that is not changeable (Spiegel, 1965).

I would not completely agree with Spiegel, et al., however. I have seen remarkably long-lasting changes occur with highly suggestible clients as a result of the extinction of maladaptive CERs. However, these clients do represent the most difficult population to work with. When highly stressed, these clients may lose contact with reality (Spiegel, et al.).

I encountered a moderately disturbed OCD client who, following a traumatic experience, developed a pseudo-psychotic state. She was initially sent to a highly respected and very expensive in patient facility, where more than one psychiatrist was convinced she was schizophrenic. However, the senior psychiatrist corroborated my conclusion that she was experiencing a pseudo-psychotic dissociated state. He did not allow her to be medicated and she spontaneously recovered. She returned to her moderately dysfunctional obsessional thinking and behavior but was no longer 'psychotic.' In everyday life, and particularly when experiencing extreme stress, this client was victimized by her exceptional trance ability.

Spiegel, et al., concludes that with somnambulists,

"... the problem is not of putting them into trance states, but of keeping them out of trance.... with the exception of occasional uses for anxiety control or abreaction, hypnosis has little place in the intensive psychotherapy of the highly hypnotizable patient. It is the identification of this profound hypnotic capacity which is critical to their therapy." (p. 327)

I found that through hypnosis, trance therapy (separation therapy), and by understanding the somnambulistic clients' cognitive style, their resistances become more understandable and less frustrating. This understanding may open the way for therapeutic change.

Hypersuggestibility usually decreases following abreactive extinction or in-vivo extinction of the CERs subserving the hypersuggestible state, but this is only true when the hypersuggestible state is maintained by those emotions. As stated, some clients are inherently somnambulistic, and decreasing their anxiety will not cause a decrease in their suggestibility. My experience indicates that, even with severely regressed clients, hypnosis is safe. I never witnessed a severe or chronic decompensation as a direct result of hypnosis. However, the combination of forced abreaction, dissociation, and flooding may cause a traumatized hypersuggestible client to withdraw into the aforementioned dissociative psychosis. This is extremely rare, and I personally have never seen this.

One client early in my career rapidly decompensated immediately after I began a hypnotic induction by asking her to just find a spot a little above eye level and gaze at that spot. She quickly regressed into a fetal position and began muttering. When she came out of this state, she appeared spacey and dissociated. This became a permanent condition. On a two year follow-up, she had remained in the same condition. She had appeared perfectly normal during the initial interview, and her presenting problems were A-B, although she was not severely emaciated. This is the only client I encountered who became suddenly psychotic with no prior history of it, and I have no suggestions as to how to handle psychotic clients using my techniques.

Understanding of the Law of Reversed Effect, also termed the Law of Reversed Action, and its clinical utility is necessary. Kroger and Fezler (1976) state:

> "... the harder one tries to do something by using his will the less chance he has of succeeding ... whenever the imagination and will come into conflict, the imagination

invariably wins. This principle is similar to that of reciprocal inhibition developed by behavior therapists."

The Law of Reversed Effect is the same as Frankl's (1960) concept of paradoxical intention in which clients are instructed to try very hard to perform the act they find distressing. Frankl treated a young physician who was plagued with inappropriate perspiring and fear of perspiring. Whenever the physician anticipated an outbreak of perspiring, the anticipatory anxiety triggered the sympathetic over-arousal that caused the excessive sweating. Frankl states: "I advised the patient, in the event that sweating should recur, to resolve deliberately to show people how much he could sweat. A week later he returned to report that... after suffering from his phobia for four years, he was able, after a single session, to free himself permanently of it" (p. 196).

The Law of Reversed Effect is an important tool in the hypnobehavioral treatment of sexual dysfunction. I explain to clients that the harder they try to force an orgasm or erection, the more the sympathetic division of the autonomic nervous system becomes active, which causes more vasoconstriction in the genitals. This prevents the orgasm or erection, and often triggers permature ejaculation or inorgasmia. The Law of Reversed Effect is important to remember when using imagery to elicit and extinguish negative feelings or to induce relaxation. Clients who try to force themselves to imagine situations or try to create the desired feelings through volition fail and become frustrated. This of course impedes the desired results.

The Law of Reversed Effect makes sense from an adaptive point of view. The harder people try to exercise their volition to decrease CERs, urges, and suggestions, the stronger these grow. The converse is also true: the harder people try to force themselves to feel fear, CERs or urges, etc., the more impossible it becomes and the more they relax. Often doing this elicits reciprocal inhibition. Through humor, clients often find it amusing to try as hard as they can to make themselves anxious and miserable.

Respondently conditioned fears are often adaptive because they cause an avoidance of stimuli that are aversive or signal events that may be detrimental. The probability of survival would be decreased if people were able to voluntarily turn off fear and not avoid dangerous situations. Conversely, if we could voluntarily turn on fear, fear could become associated with benign stimuli and mediate avoidance of situations that must be encountered in order for people to survive. When food becomes a stimulus that mediates strong avoidance, the person becomes anorexic. A-Bs are able to easily auto-condition themselves to strongly fear food.

Dissociation is the blocking out of consciousness of the memories of when and how negative CERs were acquired. How often this really occurs is controversial. However, even when dissociation occurs, the stimuli-response associations remain because people still feel fear when encountering fear-eliciting stimuli. One may not be able to consciously recall how the stimulus-fear association was acquired. Humans are able to inhibit (dissociate, deny) conscious awareness of learning situations but not the basic stimuli-response associations. What is important is that CERs do not fade away over time, but do so only through extinction. This has been demonstrated in lower animals as well as in man.

I instruct clients to use the Law of Reversed Effect to decrease the negative feelings they encounter during the abreactive extinction process during conversion reactions or whenever they experience urges to binge, vomit, starve. Successfully decreasing these CERs helps them form positive expectations that they can control their fears.

In review, most A-Bs and OCDs experience strong urges to commit unwanted acts and the act reduces the urge. These urges are elicited by covert and overt stimuli that clients may or may not be aware of. Chronic A-Bs and OCDs are usually unaware of how negative CERs were acquired and therefore feel out of control. The feelings of loss of control and the resultant anxiety increase the negative CERs' intensity that mediates more maladaptive avoidance behavior.

Hypnotic phenomena can serve as an explanatory model for compulsive behavior. The following example demonstrates how during a suggestible state people can acquire an urge to carry out a specific act, act on that urge, and then rationalize their irrational behavior. People often act, do not know why, and then try to make sense out of their actions. Various hypnotic variables, such as hypersuggestbility and spontaneous amnesia, offer an interesting model for OCD and conversion reactions.

While demonstrating hypnosis, I gave a student who had never previously been hypnotized the suggestion:

> "Upon awakening, you will feel an overwhelming urge to stand up, walk over to the window, open the window, and then you will sit down. You will feel this overwhelming urge when I look at you and touch my right index finger to the tip of my nose. You will temporarily be unable to recall this suggestion after awakening from hypnosis. I am asking you to do this in order to demonstrate to you and to the class the phenomena of repression that we have just discussed. If you agree to accept these suggestions, please nod your head."

After the student nodded his head in agreement, he was awakened. Approximately five minutes later during a discussion unrelated to hypnosis, I looked at the student and touched my right index finger to the tip of my nose. He became visibly agitated, looked at the window, fidgeted, and finally walked over to the window, opened it, and returned to his seat. When I asked why he opened the window, he replied, "It's stuffy in here. I felt we needed some fresh air." I then stated, "But it's the middle of winter, and cold in here - so why would you open the window?" The student appeared more confused, and finally admitted that he did not know why he had acted as he did.

Hypnotic suggestions and compulsive urges are similar in that they are often overwhelming and difficult to resist. The Law of Reversed Effect applies to both; the harder one tries to resist an urge or suggestion, the stronger it becomes. Anxiety most likely increases as people try to block the suggestion, thus causing the suggestion to become stronger. This is reasonable, especially if anxiety, fear of loss of control, etc., are the mediators for the urges or suggestions. A positive feedback loop often forms. The person feels anxious about the urge and the anxiety causes the urges to build. The higher the anxiety, the stronger the urge becomes to commit the act that reduces it. If the compulsive act elicits more anxiety after it causes relaxation, the urge to commit the act will again emerge. The cognitive and behavioral feedback loops that perpetuate much anxiety-mediated behavior are discussed more fully in the following chapter.

The next example demonstrates that even when subjects remember hypnotic suggestion, they may still be unable to resist following them. This supports the conclusion that insight into why we behave the way we do seldom results in a behavioral change. Knowing 'why' on the cognitive-verbal level for many people may have little effect on the behavioral level, and even less effect on emotional reactions. The following illustrates this point.

I hypnotized a student and gave the suggestion that her slacks were glued to the chair and that she would be unable to stand up upon being awakened. I also told her that when she did try to stand up, the harder she tried to stand up, the more impossible it would become. I added she would be temporarily unable to recall the suggestions. Immediately after being awakened, the student defiantly told me, "I remember the suggestions!" I replied, "Okay. Well, let's break for lunch then, since it's now after 12 noon, and everyone is probably hungry." As the rest of the class began to slowly leave the room, the subject remained sitting but commented, "Well, I can get up if I want to." I replied, "Then try." The subject finally started to try, but was unable to rise. However, she kept insisting that she could if she really wanted to and continued rationalizing that she did not want to get up. She finally became agitated and angry when she again tried and failed. As her fear and confusion increased, she began to employ a variety of coping behaviors -- first rationalization, then anger, and finally pouting. I reinstated hypnosis and removed the suggestions.

This illustrates that, even if the subjects are aware of the how's and why's of hypnotic suggestions, they may still be unable to resist performing the suggested behavior. Compulsive urges appear to function in a similar way. This reminds me of clients who, after years of insight-oriented psychotherapy, have stated, "I know exactly *why* I do what I do, but I still do it." Sitting for hours and listening to clients' ruminations and rationalizations as to how and why they think they acquired their maladaptive behavior has little therapeutic value (Eysenck, 1952; Hobbs, 1962; London, 1969).

Verbal behavior and cognitions often function separately from emotional states and motor actions. Clients often feel CERs, behave maladaptively to reduce those feelings, and then make sense out of their behavior by intellectualizing or rationalizing. Many clients become so skilled at employing their maladaptive coping behavior that the anticipation of the emotions alone can elicit immediate coping so that the emotions are not at all experienced. This prevents even slow or partial extinction of the emotions.

The aforementioned brings us to the core disagreement between behavior modification and insight-oriented therapies. Most of the time people do what they consciously intend to. However, A-Bs, ADs, and OCDs often verbalize one thing, feel another, and behave in a way that may be incongruent with their verbalizations or feelings. At the heart of this controversy is the basic question: Does changing a client's thinking (cognitions and attitudes) and stimulating insight cause changes in behavior? The answer appears to be 'No'.

Leon Festinger (1964) concluded that attitude change is "...inherently unstable and will disappear or remain isolated unless an environmental or behavioral change can be brought about to support or maintain it" (p. 416). Zimbardo and Ebbeson (1970) concluded that,

> "... changes in attitude are not necessarily accompanied by changes in behavior. Furthermore, when changes in behavior do occur they are rarely, if ever, general

or enduring. When would the social learning approach expect verbal statements to match nonverbal behavior? Essentially it would predict a match whenever a person expects similar consequences for both kinds of behavior." (p. 85)

Changing attitudes without changing behavior or expectations is obviously an inefficient means of altering one's behavior (Hobbs, et al., London, et al.). It appears that insight-oriented psychologists and educators, at least in the area of attitude change and prejudice, have based many of their methods on a false assumption (Mikulas, 1974). However, changing people's behavior often results in corresponding changes in their cognitions. Cognitions (attitudes) tend to become congruent with behavior.

For example, Bandura, Blanchard, and Ritter (1969) successfully treated snake phobics with desensitization and modeling procedures, and found that as the subjects' fear of snakes diminished, their attitude towards snakes became more favorable. I have found that following the extinction of the CERs elicited by food in A-Bs, their attitude toward food becomes one of relaxed indifference rather than fear and guilt.

Humans can experience motor actions whose origins are not known to them, yet they try to make sense of it. Delgado (1960) stimulated the internal capsule areas of a person's brain that caused his head to move. Even though he did not willfully move his head and had no knowledge of how the movement was induced, he still tried to make sense of it. When asked why he had moved his head, he stated various reasons such as, "I am looking for my slippers." Apparently, our cognitive system has to make sense of the actions of our behavioral system.

The Spiegels, et al., explain compulsive behavior by the interaction of three variables which they term 'the compulsive triad': 1) amnesia to context or origin of the signal that elicits the compulsive urge; 2) compulsive compliance with the signal; and 3) rationalization of the compliance.

In the first hypnotic demonstration, all three features of the triad are apparent. 1) The subject had amnesia as to how he acquired the stimulus-response association (feeling the urge to open the window when I touched my nose), 2) the subject had complied with the signal after he felt the mediating emotion by opening the window, and 3) he rationalized his behavior by saying that the room was stuffy.

In the second demonstration, the subject remembered the suggestions, but still felt the strong urge to comply. When she complied with the suggestion, she also rationalized her compliance. The Spiegels agree that people are often compelled to act for reasons they have forgotten and attempt to rationalize their behavior. I am in agreement with the Spiegels up to this point. However, they state that a purpose of psychotherapy is to have the client overcome the amnesia. The Spiegels are in error if they assume that the client becoming conscious of the origin of the acquired associations between the signals and the urges will result in behavioral change. As shown, insight into this association may cause little to no change in behavior.

As stated, many A-Bs and OCDs enter a trance or regressed state when they perform their compulsive rituals. During this trance, they avoid experiencing the negative mediating CERs and perform their compulsive behavior automatically. Trances and amnesia are avoidance behaviors that guard the mediating CERs and operant behavior (e.g., binge-eating and vomiting) from extinction. As stated, simply removing the amnesia may not cause the emotions or overt behavior to change. Extinction of the emotions and overt behavior has to occur in order for them to decrease. The stimuli that elicit the troublesome urges have to be identified and the urges extinguished by repeated elicitation before the compulsive urges decrease. Removal of the amnesia is, however, often helpful in identifying the eliciting stimuli that need to be presented to the client so that the extinction procedures can be enacted. Finally, new adaptive and incompatible behavior must be conditioned in the place of the old compulsive behavior following extinction. Simply extinguishing the mediating urges and not establishing a behavior that is incompatible with the compulsive behavior may only result in a short-lived decrease in the compulsive behavior and obsessions.

Bringing the rationalizations closer to the original signals (stimuli) will sometimes cause the emotions to be elicited, and partial extinction will occur. If the mediating CERs are not repeatedly extinguished, spontaneous recovery will occur. This accounts for the ineffectiveness of analytic or insight-oriented psychotherapy. As emphasized, the insight that clients gain through analysis result in behavioral change only to the degree that they experience the mediating CERs and to the degree these emotions extinguish, as well as to the degree that the expectations are altered.

As stated, hypersuggestibility is often a major variable in the origin and maintenance of behavioral problems and may result from chronic or acute anxiety. Early anxiety often sensitizes people so that later anxiety-eliciting situations mediate maladaptive behavior. Many hypnotherapists and hypnoanalysts (Honiotes, 1980; Kappas, 1978; Kroger and Fezler, 1975) believe that most ADs originate from the acceptance of maladaptive suggestions during childhood. These suggestions and expectations, unless altered, guide clients' behavior throughout their lives. Children and many suggestible adults accept suggestions either literally and/or by inference, and some of their inferences may be incorrect. For example, children who are punished may infer that they are not loved, even though their parents show their love in many ways. The hypersuggestible nurse mentioned earlier had accepted that the reason her boyfriend was aloof was because she gained weight when in reality she had not is another good example. Her incorrect conclusion caused her to starve and jeopardize her life.

Once suggestions are accepted in a hypersuggestible state, they become difficult to change through the clients' conscious efforts or through standard educational procedures. Information that is dissonant with a person's beliefs or ideas causes a state of tension that the person attempts to reduce

(Festinger, 1957). Many clients reduce this tension by simply dissociating and rejecting any information which is incongruent with the already accepted suggestions and are usually unaware that they are doing this. Most people's first reaction to cognitive or behavioral change is resistance. There is a behavioral or cognitive inertia that must be overcome in order for a person to change. This is reasonable because it is adaptable for early learning to be difficult to change. Many early learned attitudes and habits facilitate survival and should not be easily changeable. Higher levels of anxiety correlate to less tolerance for cognitive dissonance as well as tenaciously holding onto beliefs even though they may be irrational.

In order for maladaptive suggestions to be altered, clients must be put in a hypersuggestible state similar to the one they were in when they originally accepted the suggestions. Through hypnosis, most clients can be regressed to this state; while in the hypnotic state, clients often spontaneously exhibit regressed behavior and cognitions. The similarity of the thinking of regressed or suggestible adults to the thinking of a child explains how they can accept their own inferences or others' literal suggestions as the absolute truth and govern their behavior by them. Some clients appear to be cognitively fixated at a childlike state and, like children, lack critical thinking, are generally hypersuggestible, and rely heavily on trance logic.

Critical thinking is learned and as a result people become less suggestible as they learn more critical reasoning skills. Research supports the view that children are more imaginative and suggestible than adults (Messerschmidt, 1933a, 1933b; Stukat, 1958). Hypnotic susceptibility increases from ages five to between nine and fourteen (London, 1965; Barber and Calverley, et al.). Barber showed that post-hypnotic amnesia was demonstrated by 80% of children tested between eight and ten years of age, but that it was much more difficult to produce in 18 to 22 year old subjects. Older subjects were also less proficient at producing hallucinations. It is normal for children to function in a continual state of hypersugggestibility, but not for adults. Also, it is not considered abnormal for imaginative children to hallucinate.

Trance logic is best viewed in children. Children may assume that because an old woman has a long crooked nose she must be a witch, and may then act on this assumption by annoying her. When we consider similar examples from our own childhoods, it is easy to understand how even a normal imaginative children who are not traumatized may accept false suggestions and govern their lives accordingly. Suggestions and the subsequent conclusions drawn from them easily become unconscious as a result of the spontaneous amnesia experienced by children.

Therefore as children mature, they find themselves behaving in ways or experiencing feelings and thoughts they cannot explain. Amnesia for the origin of these behaviors creates fear of loss of control and anxiety and may increase clients' suggestibility and tendency to uncritically accept more suggestions. Supporting the hypothesis that anxiety increases suggestibility is the finding by Shapiro (1971) that anxiety predisposes patients to placebo effects. It has been reliably documented that a hypnotic suggestion may survive time-dependent decay (spontaneous decay) for many years, even in adults.

M. Bulimic Cognitive-Behavioral Feedback Loop

The basis of my explanation of how compulsive-addictive behavior not involving physiological addiction becomes established is that the out-of-control consummatory behavior clients are addicted to must initially elicit relaxation and then unpleasant emotions. This is essential for the formation of addictive-compulsive behaviors such as bulimia and may also be a significant maintaining variable in drug addiction as well.

Eating is primarily reinforcing and food becomes a secondary reinforcer because it has been associated with love. Therefore, when bulimics experience negative emotions, they eat in order to reduce these feelings. As with anorexics, bulimics are usually somnambulistic and easily enter dissociative states during binge-eating which blunts the negative feelings. During binge-eating, most bulimics and many chronically obese clients regress to a childlike state where they feel more secure. The time distortion experienced during these dissociative states is reinforcing because it reduces negative emotions just as time distortion can be used to reduce pain.

For A-Bs, food becomes associated with rejection (separation anxiety) usually during childhood and pre-adolescence when they are taught or conclude being fat means they are worthless, etc. Hypnotic regressions and the initial interview usually reveal these fears were reinforced by teasing if they were overweight and by their parents' and peers' behavior and attitudes toward obese people. The fears elicited by weight gain -- or even the thought of weight gain-- motivates the A-Bs' periods of starvation.

Secondary gains may enter the picture. Binge-eating, vomiting, or starving are often reinforced when used as a weapon against the parents. I have encountered cases where the mediating emotions were relatively weak in comparison to the reinforcement gained by being able to control and punish a significant other.

Another way to elucidate the basic association between weight gain and separation anxiety and the association between food and love is through word association and sentence completions. This may be done when clients are in hypnosis. The therapist should instruct them to give their first response to the stimulus word without analyzing their response. A list of unimportant words with the important words interspersed is then presented. Long latency responses or no response usually indicates that the stimulus word elicited negative affect and the associated word or feeling is being blocked. Clients often respond to the word *fat* with 'hate,' 'fear,' 'unloved,' or 'rejection.' The word *food* may elicit responses such as 'fear,' 'love' 'relaxation,' or 'security.' The associations can also be elucidated by regressing the client back to the situations where she accepted these associations. There are a variety of situations that reinforce the associations between food and love, such as food being used as a reward and being paired with holidays. At the same time, the popular media and current fashion styles and models suggest that only skinny women are beautiful. The influence of European designer fashion has been recognized since the 1840s, when Henry David Thoreau wrote, "The head monkey at Paris puts on a traveler's cap and all the monkeys in America do the same" (1906).

AD clients in general and OCDs in particular exhibit a variety of maladaptive cognitive feedback loops. For example, I encountered a client whose major complaint was depression and sleep difficulties. The more the client worried about sleeping, the more impossible it became for him to initiate and maintain sleep. The Law of Reversed Effect was operating against the client. The client incorrectly believed that if he had a bad night's sleep, he would not be able to function the following day. He was obsessed with the idea that, "If I don't sleep well, I will fail; if I fail, I am no good because I am the sum total of my achievements. So, if I do not achieve, I am nothing." This became a self-fulfilling prophecy, as he would find it more difficult to sleep because he was worrying about not sleeping and then would make mistakes the next day.

His maladaptive thinking was based upon strongly entrenched irrational beliefs and a maladaptive value system. First, the assumption that if he has one bad night's sleep he will not be able to function well is not true. Research has shown that one bad night's sleep may have little to do with the following days' physical or mental functioning. The second incorrect belief is that his personal worth is determined by the sum total of his achievements. This causes chronic insecurity because unless he is always a 'winner,' he feels worthless. He feels in continual jeopardy because of the fear that sooner or later others may realize that he is not perfect. He cannot form an open loving relationship with anyone because of fear that if anyone really gets to know him, they will discover his weaknesses and not like him. Therefore, he unconsciously sabotages relationships when he feels another person is 'getting too close.'

Often chronic AD clients do not want to grow up and accept adult responsibilities. This attitude may be based upon the belief that to grow up and become independent means that their parents will withdraw their love. Also, after observing the attitudes and lifestyles of significant adults close to them, they may have concluded that being a responsible adult has little to offer. Clients should be made to see that there are more rewards for behaving in an adult manner and accepting adult responsibilities than by remaining a child. They must be convinced that it pays for them to accept responsibility and that parental love does not decrease as a function of the physical distance between them and the home. They must understand that living as a well-integrated adult does not mean that we abandon our childlike desires to have fun. Mature adults lead balanced lives by working hard, achieving, having fun, and most importantly loving and accepting themselves.

In order for clients to accept themselves, they have to confront themselves honestly. When most people confront their physical, emotional, and intellectual self, anxiety is experienced (Gur &

Sackeim, 1974, 1978). Most A-Bs and ADs are phobic of themselves and are afraid to look at themselves objectively because they are discontented with who they are. Gur, et al., have shown that self-confrontation even in ADs elicits increased autonomic arousal and was experienced as aversive. Clients who have a phobia of the self usually avoid self-confrontation by dissociating those aspects of the self that they find unacceptable. If this dissociation of important aspects of the self becomes extreme, Dissociative Identity Disorder may result.

Usually there are two fragmented personality parts that subserve clients' A-B behavior -- a younger part who uses food as a substitute for love controls the bingeing, and an older part (usually an adolescent) who fears weight gain and controls the vomiting, starving, and laxative and diuretic abuse. A-Bs may also become obsessed with repeatedly forcing their parents and significant others to prove that they love them. At the same time, a fragmented part is binge-eating, vomiting, or starving in order to punish the others for not giving them the love that they want. However, because 'good little girls' do not punish the people who love them, they feel guilty. The guilt in turn motivates self-punishment and even suicide through starvation. The A-Bs' behaviors cause significant others to become frustrated, angry, and frightened. These reactions may be interpreted as further rejection, and therefore intensifies their struggle with these significant others.

A vicious self-perpetuating feedback loop is formed. The A-B is angry at significant others and punishes them by her self-destructive behavior and hates herself for doing so. The self-hatred and guilt are dissociated and cause her to cope through more binge-eating, vomiting, or starving. The self-destructive behavior punishes both herself, which reduces her guilt, and the others, which satisfies her desires for revenge.

Maladaptive and unreasonable expectations often cause AD clients to perform maladaptive behaviors, and this behavior often causes reactions by others which are interpreted by the clients as support for their false cognitions. Once this vicious feedback loop is formed, it feeds upon itself as negative expectations help generate negative consequences. Maladaptive thinking and behavior therefore becomes strengthened and automatic.

For example, when people feel that most others do not like them, they may react in a defensive manner that causes people to dislike them. This confirms their initially incorrect supposition that people don't like them. People tend to receive what they expect from life and tend to be treated in the way they expect or allow others to treat them. Because many clients believe that attention of any kind is love, they employ maladaptive behavior in order to receive it. Again, these incorrect cognitions have to be changed. I have encountered A-Bs who admitted that they would imagine themselves in a hospital room with all their loved ones clustered around them, worrying about them. This attention-getting expectation has to be exposed and must not be reinforced.

Much emotional blackmail and passive aggression is present in chronic A-Bs, ADs, and depressives. For example, a 37-year-old A-B was punishing her husband by destroying her physical attractiveness because she was angry at him for rejecting her. She felt that if she ended up in the hospital, he would feel guilty. However, the more emaciated she became, the more he withdrew, and the more desperately she binged, vomited, and starved. When she was made aware of her motives and became conscious of the part of the personality who was promiscuous and had an affair in order to punish the husband, she expressed a desire to see a female therapist. I explained the situation to the new therapist, who did not choose to continue along the same line of therapy. Three months later, I called to see how she was. She reported at first that she was doing better, but then admitted that she had slashed her wrists a few weeks earlier. This particular client was unwilling to confront her motives and the game that she was playing, and did not improve. I have found that clients designated as borderline are not good candidates for my therapy.

Dichotomous thinking (the idea that the world works on an all-or-none basis) is maladaptive and characteristic of rigid, compulsive and perfectionistic people. For example, a bulimic who was binge-eating and vomiting five times per day before treatment, but afterwards binges and vomits once per month, may continue to hate herself as much as before. She may feel that she must be perfect, which in her mind means that even one slip ruins all that she has accomplished. These incorrect ideas about how the world and human behavior function must be changed.

I explain to clients that everything is on a continuum, and there are few absolute dichotomies. They must understand that A-B consists of behavior patterns that are learned and could be learned by almost anyone in the right circumstances. Therefore, if they occasionally slip back into the bulimic pattern, it does not mean that they are 'sick' and will eventually be back where they started. Rather it is the degree to which a person is involved in maladaptive behavior that determines whether they should be labeled A-B, AD, psychotic, etc. The disease model often reinforces the idea that a person is either mentally ill or healthy. Some A-Bs compare themselves to alcoholics and apply the Alcoholics Anonymous philosophy to themselves: once a bulimic, always a bulimic, and one slip means that they are totally 'sick' once again. Of course this expectation is often followed by the feared behavior and the ensuing thought, "I blew it so I may as well go all the way."

Many ADs have simply modeled their thinking and behavior after their parents and mainly their peers (Harris, 2009). The incorrect cognitions that often determine clients' behavior can be made conscious and altered through hypnotic regression separation therapy and direct suggestions. As stated previously, understanding the motives of the regressed child-like part of the personality may help eliminate resistances which block therapy. Also, abreactive extinction and flooding should be used to extinguish the maladaptive-obsessive thoughts and negative emotions. Hypnotic suggestion and imagery conditioning exercises should also be employed to implant positive feelings and expectations and can be reinforced by having clients use hypnotic induction tapes at home.

The negative programming from the parents and/or society has to be altered in order for permanent changes to occur. Repetition of the positive suggestions in as deep of a state as possible is best. When clients experience amnesia for hypnotic suggestions, it usually indicates that they are being accepted.

N. Spiritual Aspects of Psychotherapy

Clients may possess spiritual beliefs that are important or even central to their maladaptive behavior. Many A-Bs and OCD clients maintain the belief that their spiritual worth or the meaning in their existence is a direct function of their 'good' behavior and whether or not others like them. These maladaptive beliefs, prevalent children and among adolescents, are usually acquired during interactions with significant adults and peer models. Many parents of OCDs and A-Bs are typically rigid, aggressively overprotective, and unable to consider their child as a person in her own right. Hypercritical parents cause their children to become hypercritical of their own selves. However, the evidence supporting this is correlational. It could well be that OCDs and A-B have strong genetic causal elements. Also I have encountered clients who apparently were reared in the best possible home. However, most of these clients appeared to have experienced difficulty in a variety of areas from early childhood and may have an organically-based personality disorder. Also the role of the peer group may be more powerful in determining young people's values than their parents (Harris, 2009).

Many clients believe that their parents', peers', and, if they are religious, God's love is conditional dependent on their behavior. As stated, I contend that the core anxiety in most ADs and in particular A-Bs can be traced to the feeling that one is not loved or that parental love is conditional upon living up to what is believed are the parents' and peers' expectations. Unconditional love and acceptance from at least one significant caretaker is the major variable in the development of high self-esteem, unconditional self-love and security. Without this, most people develop emotional problems mediated by separation anxiety.

Religious indoctrination begins early for most people and can set the stage for self-hatred, guilt and low self-esteem. One need only to view the teachings of the Catholic church and fundamental

Christians concerning sex and sin to see how children may come to view their natural desires as sinful and God as unforgiving and unloving. The Christian message is either not taught correctly or simply not understood by the vast majority of people in the Judeo-Christian society who label themselves as religious.

The Judeo-Christian teachings may be purposely misrepresented by parents and various social institutions in order to control the child or more commonly the significant authority figures are simply ignorant of the Biblical teachings concerning self-love, guilt and sin. Many churches and Sunday schools often reinforce clients' incorrect beliefs through both direct and indirect suggestions. Some ministers, particularly the lay ministers and Catholic brothers, are ignorant of or misinformed about the important Christian teachings that guard against the development of much self-destructive, anxiety-based behavior.

It is essential that religious misconceptions be corrected, especially with clients who exhibit guilt and perfectionism. At the beginning of therapy, clients often insist that their religious upbringing has nothing to do with their current emotions and maladaptive behavior. When resistance to the hypnotic suggestions for self-love occurs, however, this may indicate that a dissociated and regressed part of the personality is resisting because those suggestions are incongruent with its cognitive set. Through hypnotic regression, separation therapy and direct suggestion, the old programming can usually be altered.

For example, a successful client experienced extreme guilt concerning an extramarital affair but insisted that her religious upbringing had nothing to do with her current problems. I suspected that her religious upbringing was causing a problem, so I assigned certain Bible passages to be read as homework. She reported experiencing anxiety attacks when she attempted to read any of the assigned passages, or even when she touched the Bible. She was highly motivated, and followed through with systematically desensitizing herself. She first listened to religious tapes explaining the Christian message concerning guilt, self punishment, etc., and also read other sources about the Christian doctrine concerning love and guilt.

Only after she extinguished her anxiety did the positive suggestions take hold. Her anxiety had caused her to dissociate the hypnotic suggestions concerning self-love and self-worth given at the start of therapy. Therapy was a slow process that extended over a period of six months, at which point she rated herself as 90% improved. I must emphasize again that the guilt seldom disappears following MHPs' explanations of the Christian doctrine concerning guilt and self-love, but a better understanding of these concepts can often reduce those resistances that are reinforced by the desire for self-punishment. Circumventing the critical area of the mind through hypnosis is much superior to simply talking to the analytical mind.

Once clients understand the importance of self-love, the limited value of guilt, and that the MHP is not judging them, communication with the client is facilitated. This decrease in resistance enables the guilt and associated negative ideas to be more easily identified and extinguished. After extinction has been accomplished, the positive hypnotic suggestions explaining the Christian doctrine are much more easily accepted.

If MHPs are uncomfortable in dealing with religious misconceptions, they should seek the aid of a competent religious counselor. When treating clients who have been reared in an authoritarian dogmatic church, a priest or minister from that particular denomination should be employed, as many of these clients are suggestible only to the 'real thing,' which of course the MHP is not. As previously stated, many clients become selectively attentive (suggestible) to a few significant individuals. Those who are authority figures early in life have the most powerful influence. There are many knowledgeable religious counselors in all denominations who understand the Christian doctrines and their assistance should be utilized when these clients are encountered.

For many clients, this aspect of therapy cannot be treated lightly, and if omitted, the therapy most likely will fail. Most MHPs do not effectively aid their clients in resolving these important conflicts, and have not studied the religious doctrines accepted by most. Out of respect for my clients, I have read the Bible, the Book of Mormon, and most of the Koran as well as associated study materials.

The following is a typical explanation that I give to clients concerning the Christian doctrine of love, guilt, and sin. Clients are additionally encouraged to read Romans I through IX in the modern English version of the New Testament. If they prefer, I read the relevant sections to them and/or record it for the client to listen to. These suggestions, particularly those concerning self-love, should be given under the deepest hypnotic state that can be induced. The following is a transcript of the explanation and suggestions that I give:

> "Four hundred years before the Commandments were given, God entered into a contract with Abraham. He promised Abraham that he and all of his descendants would inherit the Earth, not because Abraham had done everything right, but because Abraham loved and had faith in his Heavenly Father.
> Four hundred years after the contract between Abraham and God, which was based on Abraham's faith and love of God and not upon Abraham's behavior, God gave the Jews the Ten Commandments in order to bring humans back to faith. Therefore the Commandments were not given to them to condemn them to eternal Hell, but to bring them back to faith.
> God's message throughout the Bible is that He loves His creation, mankind, and that His love is unconditional. Through creation, we are imperfect, and only God is perfect; therefore, everyone will commit sins because all of us are imperfect. If everyone was condemned to Hell for violating the Commandments, no one would be able to enter Heaven, and God's promise of Heaven would be worthless.
> It must be kept in mind that God's promise to Abraham, who is the father of us all, was based on faith; neither Abraham nor God changed the contract. God does not go back on His promises. When a Christian violates a Commandment, they are supposed to ask God for forgiveness because they have faith in what God has promised, and then one is forgiven for all eternity. The only sin a Christian can commit is to lack faith; but again, if one believes in God and has faith in His Word, then one should ask God for forgiveness. Forgiveness is then given for all eternity.
> God demonstrated His unconditional love for us by sacrificing His only begotten

Son. The death of Christ symbolizes God's love for us, because even after man killed His only Son, God still loved us. In the New Testament, Paul explains the Christian Doctrine concerning sin and guilt in his letters to the Romans. He states that if God's promise was based upon our adhering to the Commandments, the no one would enter Heaven because no one can always uphold the Commandments. Therefore, again it is only faith in our Lord, Jesus Christ, that we can attain everlasting life.

God is the Judge and the Punisher; we do not have the right to judge nor punish ourselves, nor to judge and punish others. This is God's right, and His right alone. In other words, God has taken that responsibility from us -- in fact, it was never ours. The only sin we can commit is to lack faith and to not love our Heavenly Father. By judging ourselves and punishing ourselves and others we are putting ourselves on the plane of God. This is of course a sin because God said that there is only one God.

Prolonged guilt is a sin, because when we commit a wrong act or think a wrong thought, and we ask for forgiveness, we are forgiven for all eternity. If a Christian has faith and believes in what God has said, there is no room for guilt in our lives. Again, we are imperfect, therefore we will sin, and of course we will be forgiven by our heavenly Father Whose love is unconditional. Paul, in his letters to the Romans, said that nothing can separate us from God's love. We do not have the right to judge ourselves, punish ourselves, or judge and punish others. That is God's right, and His right alone.

The things that you must do according to the Christian Doctrine are: you must love God with all your heart and love your neighbor as you love yourself. This is what Christ replied when asked what Commandments we should follow. According to the Christian teachings, this is what we should strive for every day.

This is an excellent philosophy of living for you as a Christian. When you violate a commandment, you are sinning. If you continue to feel guilty or hate yourself after you have asked for forgiveness, you are sinning again, because by doing this you have judged and punished yourself. The only thing you can do as a Christian is to accept the Word of God and ask for forgiveness, love yourself, and treat yourself with respect and do the same to others. These few rights that God gave you are good ones, because they mean that any time you are hurting yourself or another, you are doing wrong; but if you are loving yourself and your fellow man, and appreciating the gift of life, you are following God's plan. The Christian philosophy of life is practical and leads to a fuller and much happier life."

Once these teachings are explained, the negative self-statements and the associated emotions can be flooded out or extinguished by having the clients list their negative cognitions and then ruminate on them until the negative thoughts and emotions are impossible to emit. The same principles concerning flooding and extinction as applied to other negative thoughts and emotions in prior chapters apply.

The Christian doctrine concerning self-love and how we should treat ourselves and others is emotionally rewarding for the individual as well as practical. Most atheists understand and respect that following certain rules of living causes people to feel better, be spiritually (emotionally) freer, and to enjoy life more and lead fulfilling lives. Clients must realize that it is only by helping others and treating themselves and others with love and respect and, above all, by appreciating the gift of life, will they succeed in having a meaningful life.

The Christian God should be depicted as the perfect parent, an all-knowing unconditionally loving, and always forgiving Deity. It is possible with religious clients and those who believe in some concept of God to have them substitute God's love for their parents' love. Clients must understand that all parents, because they are human, are imperfect. Some parents are incapable of loving anyone, including their own children. For clients to hope that their parents will love them or show them the love they require, and in the exact way that they think they require it, is unreasonable. Emotional maturity involves accepting others as they are and not trying to make others change to suit them. The goal of my therapy is to have clients learn an adaptive set of coping skills and attitudes that help them effectively deal with their emotions, behavior, and social and physical environment.

I explain that families often assign roles to family members. These labels and expectations are not easily changed, particularly in families where members gain by keeping the individual locked into that role. For example, if a child was undisciplined, he may be perceived as being undisciplined by his family throughout his entire life. Children tend to follow the expectations of their families whether those expectations are good or bad. A-Bs usually are assigned the role of the 'perfect child,' although this is not always the case.

The following is an example of a bulimic who was assigned the role of being a 'loser' by her family. Her older sister was more achievement-oriented and more highly dependent upon the parents for emotional support. The bulimic daughter was doing her best to become independent of her family but felt extreme guilt because she had not lived up to her parents' social, academic, and physical expectations. As a result of this guilt and dissociated anger toward her parents, she followed the typical A-B pattern: she was overly concerned about her appearance, could not tolerate rejection, and was binge-eating and vomiting several times per day.

As this client was troubled by extremely low self-esteem, flooding was employed in order to extinguish the emotions mediating her negative self-statements and negative obsessive thoughts about herself. This procedure was carried out following in-vivo FRP to extinguish the urges to binge-eat and vomit and was conducted on two separate occasions. She was asked to list all of her negative self-labels and to then read them aloud. While reading them, she was coaxed into feeling all the associated emotions and to allow them to reach their highest intensity. The flooding was always continued until the emotions were zero and the client felt that the thoughts and the whole situation were foolish and untrue.

The first session of flooding took approximately two hours; the second lasted approximately one hour. During the second flooding, the obsessive thoughts and emotions extinguished at a quicker rate, following a typical extinction curve. She experienced a significant decrease in her binge-eating and vomiting following both sessions.

If parents have been unloving, clients can be persuaded to accept that fact and led to understand that it was not because they were undeserving of love, but that the parents were inadequate and missed much by not being loving parents. Clients who believe in God can be told that all parents' love is imperfect, and that even if no one in the whole world loved them, God does and always will. It should be stressed that God's love is much greater than human love, and that all individuals are unique by virtue of their creation and therefore special to God. This concept can be supported by explaining that every individual is unique genetically.

At this point, I usually describe the various gifts that the client possesses and how fortunate they really are. The laws for successful living (covered in the next chapter) are explained and stressed. These concepts are reinforced by requiring the client to listen to the recorded hypnotic suggestions made during the sessions before going to sleep. Clients also record the in-vivo FRP exercises they perform on their own. Requiring clients to do this forces them to properly execute the exercise and allows the MHP to monitor the process.

I personally like working with A-B and AD clients because the vast majority are not characterologically disordered and are not criminals. Binge-eating, vomiting, and experiencing anxiety are not good, but they are not morally reprehensible behaviors. I express this attitude often during therapy.

O. Laws of Successful Living

There are specific behavioral laws, cognitions (attitudes) and behaviors that are instrumental in people achieving a secure, fulfilled, and happy lifestyle. The possession of these attitudes expectations and behavior are causative factors for the achievement of personal, monetary, social and professional goals.

There are many definitions of success and successful living. One definition states that people are living successfully when they are progressing toward worthy goals. This definition is incomplete, but it does emphasize that goal-directed or purposeful behavior is an important ingredient in successful living and adds meaning to one's life.

However, this definition is lacking in that it does not emphasize self-love. Achieving goals, making more money and attaining prestige are all too often pursued mainly in order to enhance people's self-esteem and validate their self-worth rather than for the intrinsic enjoyment of performing the activities. Of course, certain activities when necessary must be performed, such as working at an unrewarding job when in financial distress. However, feeling satisfied with one's self is the goal of most people's strivings and represents the primary reinforcer and motivator for most activities.

A more complete definition of success is that people are living successfully when they appreciate and love themselves. This is the goal that most attempt to achieve through the acquisition of more material goods, fame, education and social status, and it explains why there is so much self-hatred and frustration in our society. Western society is attempting to gain self-love through non-spiritual means, and for most it is not working. This is obvious when one listens to clients report that they have everything (affluence, prestige) but still feel a lack of meaning in their lives. These people have found that even when these goals are achieved, they are not the primary satisfiers they were thought to be. People can feel successful and experience a sense of meaning in their lives without having achieved any of the aforementioned 'standards' of success.

Unhappy and insecure people are often confused as to what is valuable or have internalized a value system that is emotionally and physically unadaptable. For many, fluctuations in their social status, income, symbols of prestige, etc., are the sole determinants of their self-love and self-esteem. These people are only as secure as the number of college degrees they possess, their job title, their affluence and social standing, their golf score, or how thin they are. As there is always the possibility that these variables will change, the people are chronically insecure and easily threatened. This in turn generates a chronic state of anxious anticipation and worry. When they meet someone who has more of the valued items, they feel defeated and inadequate. In order to validate or increase their self-worth, they become obsessed with acquiring more symbols of 'success.' An individual who values his peace of mind, life, health, and relationships with others that are based on unconditional love and acceptance will be disturbed by fluctuations of these external variables of course, but will not become severely depressed by them. The person who values the things that cannot be easily taken is in a more secure position.

An example of someone who was successful (enjoyed his life) without achieving affluence or any of the normal societal standards of success was Henry David Thoreau. His philosophy is expressed in Walden Pond and should be read by those experiencing existential anxiety. Thoreau's philosophy helps clients realize that life can be materially very simple and still be meaningful and enjoyable. Thoreau of course did not intend his lifestyle to be for everyone, but he does explain how he found meaning in his existence and spiritual fulfillment without the achievement and accomplishments that most Americans, then and now, believe are so necessary for a meaningful existence.

A significant proportion of Western society is attempting to fulfill the need for self-love and self-acceptance through the achievement of goals which may only be achievable by a few -- or involve such sacrifice that much of the meaning in life is lost in the pursuit. In a spiritual sense, these people are worshipping the wrong god; therefore when these goals are reached, they often do not satisfy people's need for self-acceptance and love. I am not saying that achieving fame and material affluence are undesirable; on the contrary, they obviously enhance one's life. However, they are neither necessary nor sufficient for a meaningful and rewarding existence. As a result, many are left in an existential vacuum when their goals have been achieved, but they still do not like themselves or are still insecure.

The sense of meaninglessness that clients often describe is a result of not deriving pleasure or positive reinforcement from their lifestyles. In their single-minded pursuit of goals, either the ability to enjoy themselves has been lost or has never been developed. The value system that caused them to pursue goals which do not satisfy basic emotional or, if religious, 'spiritual' needs is often at the core of AD and depression in particular.

There are emotional needs that if not satisfied may jeopardize individuals' physical survival. If these needs are not satisfied, people experience a sense of loss of meaning that may cause biochemical and physiological changes in their bodies which set the stage for premature death (Seligman, 1978). This loss of meaning may also lead to depression and possibly suicide.

The following are the laws of successful living that I explain to each client. Understanding, accepting and living according to these laws will help them achieve self-love, personal, financial and professional goals and, therefore, a more meaningful lifestyle. These laws should be referred to throughout therapy as well as recorded for clients to review periodically at home. Some of these statements will possibly be direct confrontations of some of the most cherished beliefs of many clients and, therefore, may elicit strong resistance, anxiety, and self-hatred. In order to lessen the sting, I cite examples from my own life where I've paid the price for violating certain laws. Additionally, one should use some humor when confronting a client's games, immature thinking and behaviors; humor reciprocally inhibits anxiety and increases rapport. The reader must keep in mind that to be confronted and to face one's self is often an anxiety-eliciting experience.

1. Take care of your health, because the most important thing is that you are alive. If you die, you have robbed yourself of the opportunities for learning the requisite attitudes and skills for having a good life. We all have spiritual needs which can only be satisfied by appreciating the gift of life and by accepting and loving yourself.

2. The physiological facts concerning your brain and body must be faced realistically. The human brain has more areas that have to do with feeling good, intellectual curiosity and learning, artistic appreciation, laughter, and love and sex, than it has areas that are involved with negative emotions and thoughts. Your brain is, therefore, wired for you to feel more good than bad feelings. Activating these areas and enjoying life is necessary for health and is the natural state. Simply forcing a smile causes positive neurochemical changes in the brain. How good you feel physically influences how good you feel emotionally; the converse is also true.

3. Knowledge and its application is power. Our greatest enemies are ignorance and procrastination (not applying one's knowledge). The first step in gaining control over ourselves and therefore our lives is the acquisition of knowledge. The second step, and often the hardest, is to convert that knowledge into action.

4. As a general rule, you get out of life what you put into it. If you expect something for nothing, you will be greatly disappointed.

5. All behavior follows a person's expectations, self-image, and self-concept. These expectancies and self-images that guide our behavior are implanted into our minds through our imagination, not

willpower. The imagination is usually more powerful than the willpower. When the imagination is pitted against the willpower, the imagination usually wins. The unconscious mind is programmed through the imagination, not through the willpower. Because the unconscious mind is a reservoir of conditioned emotional responses and overly learned behavior patterns, it is often not directly influenced by willful choice, but does often determine what choices are made.

6. The Law of Reversed Effect (Reversed Action) states that the harder we try to use our volition to force ourselves to do something natural, such as to relax or to feel good, the more impossible that behavior becomes. For example, the harder we try not to worry, the more we worry; the reverse is also true.

This law often applies to hypnotic suggestions (unconscious suggestions and expectations). The harder one tries to use one's willpower to change or block an unconsciously mediated suggestion or conditioned emotional response (urge, feeling, etc.), the stronger that response becomes. The converse is also true: the harder one tries to create the conditioned emotional response, the more impossible it becomes. This law also applies to cognitive behavior: the harder one tries not to think about something, the harder it becomes to not do so. This law is used to rid people of sabotaging emotions and thoughts.

7. Life is boring to boring people, and life is uninteresting to uninteresting people. The converse of this is also true: life is exciting to exciting people, and life is interesting to interested and interesting people. If you find life boring and uninteresting, it simply means that you are boring and uninterested. However, you can become an interested person through gaining knowledge about yourself and the world around you by becoming involved in interesting activities and by following these laws of successful living.

8. A strong person cannot make a weak person strong but a weak person can become strong through

hard work. In other words, becoming strong in any area of your life is up to you as no one else can do it for you.

9. The easy way out of a problem is seldom the best way. This law applies to not only the achievement of professional success, but also the extinction of negative emotions and behavior. Personal and professional growth involves accepting challenges and facing fears and difficult situations instead of running from them. This is often painful and involves work and sacrifice, but is usually the only way that negative emotions, thoughts and behavior that retard development can be eliminated. Working through problems teaches the individual adaptive coping skills. These coping skills enable one to face and conquer problems and obstacles which hold back the majority of people who want 'success' but never attain it. The acceptance of challenges and responsibilities along with their resulting rewards far outweighs the consequences of avoidance (quitting) behavior. Assertive attitudes and actions are usually superior to withdrawal and unassertive behavior.

10. Mistakes are learning experiences. The only person who cannot make a mistake is a dead person. The only living person who is not in the position to make a mistake is a person who has shirked all responsibilities and is living an impoverished life. This lifestyle of course is considered to be a mistake by most.

11. The most important thing in life is to love and accept yourself. Remember the old adage, "Sticks and stones may break my bones, but names can never hurt me." Real freedom is obtained when one's self-esteem and self-love are not determined by the actions or opinions of others. Although it is nice to have others like you, to allow the fear of others' opinions to govern your life is to allow others to control you. When self-love is attained, others no longer have the power to poison your thinking with negative thoughts or to disrupt your good mental state by causing negative emotions. Other people's opinions of you are less important than the opinion you have of yourself,

especially when others attempt to undermine your self-love or achievement of worthy goals. If someone jeopardizes your or your loved ones' physical integrity, you must also realize that you and your loved ones are the most valuable.

12.   Do not listen to people who are unhappy with their lives and have not learned the skills to live successfully or to achieve their goals. They have little wisdom to offer you. The converse is also true: listen to the people who love themselves and have achieved or are achieving the style of life you desire. These individuals have obviously learned skills and attitudes which are instrumental in the achievement of the lifestyle you desire. Upon examination of the lives of most successful people, it becomes obvious that they have lived by the ideas presented in this chapter.

13.   True friends are those who love themselves enough so that they can love you -- remember that we cannot give what we do not have. A true friend would never want to see you unhappy, but wants you to get the most out of life and is happy for your successes. If you have only one or two true friends, you are infinitely wealthy.

14.   If you have someone cornered in an argument, give him a way out so that he may save face. As a result, both of you benefit. To do otherwise is a waste of time and causes negative feelings which spread like cancer. Do not 'beat a dead horse' or become involved in struggles that will end in pyrrhic victories. For example, many people waste so much energy and time fighting with others in order to win meaningless arguments or prove a point that they end up cheating themselves of time and energy which could be put to better use. Always assess what is to be gained by the battle, because we may be unaware of what we are fighting for in actuality. If a battle needs to be fought in order to obtain your goals or to protect yourself or loved ones, then you must of course fight as hard as possible.

15.   Learn to communicate your feelings clearly. People are not mind readers, so people must make their feelings and desires known. Many of the problems we have with others are due to lack of

communication. The core problem in many failing marriages and relationships is that each party is guessing how the other is thinking, and each acts on what are often incorrect inferences.

When arguing with loved ones, begin by stating the basic fact that you love them, and that what you have to say concerns their actions, but has nothing to do with the love you have for them. Criticisms and discussions that start out in this way and continue with this premise in mind seldom end up in serious arguments.

16. Everyone wants to feel loved and valuable. Therefore treat people in a loving kind way and you will have a better chance of getting them to go along with you or at least not get in your way.

17. Behaviors and thoughts are controllable by rewards and punishments; how others behave toward you is often a result of what you have reinforced.

Positive reinforcers exist in many forms such as smiling, nodding in agreement, giving compliments, paying attention to what another is saying, payment of money, or any object or thing which causes a positive feeling. Be aware of what you are reinforcing; giving people something for nothing and 'babying' them along often reinforces indolent, non-productive behavior and withdrawal from adult responsibilities. Maladaptive coping behavior which has been overly learned is what defines and describes AD, addictive behaviors, and sexually deviant behavior.

Mentally disordered people have not learned appropriate adaptive coping skills and instead rely on maladaptive coping behaviors as the solution to life's challenges. Therefore, if you are sincerely concerned with others' welfare, you will not reinforce their self-destructive behavior and attitudes. Life is not without its pain and challenges, and will not be meaningful unless those pains and challenges are met and overcome. If it is easy to do, it is rarely worth much.

18. Honesty is the best policy. If you are honest and fair in your dealing with others and yourself, you will have much less to fear. Whether or not you are religious, follow the Ten Commandments as best

as you can, as they are good basic laws for living. People who do not will ultimately have more problems than those who do. We often create our own Hell by our actions and attitudes, and to a considerable degree you get out of life what you put into it. However, do not take this too far. Bad luck does happen. You are not responsible for bad breaks or for what others do or fail to do that affects you.

19. Successful people do not have fewer problems than unsuccessful people. In fact, as they achieve more, the may encounter more problems. The difference between successful and unsuccessful people is that successful people solve their problems. They view problems as challenges and derive a great deal of satisfaction from solving them. Success is for the most part earned and not due to 'just' luck.

20. Money is important. However, we only need a certain amount of money in order to live well. The acquisition of money for the sake of money or for the acquisition of power constitutes worshiping the wrong god and will help create one's own hell. You cannot beat money for what it can buy, but if you lose the things that money cannot buy in the process of obtaining it, such as health, happiness, self-respect or life, then the price has been too high.

21. Time is life, and the most important resource you have. Wasting time is any time you are thinking negative thoughts, experiencing negative feelings, or engaged in fruitless discussions or activities which you do not enjoy or that do not facilitate your progress in obtaining your goals. The converse is also true: any time you are enjoying yourself yourself, feeling good, laughing, feeling a sense of meaning and fulfillment, you are spending your time well.

Keep track of your time so that you become aware of how you spend this valuable resource. Eliminate those activities, relationships etc., that steal your time. Keeping a log of your daily activities is extremely helpful.

22. Set short-term (6 months) and long-term (5 years) personal and professional goals. Take time to

plan your life. Formulate and write down exactly what you want, including money, family, desired residence, type of work, etc. Then formulate the steps that are needed to achieve these goals.

If you are already involved in a profession or business, work to learn everything you can about it and become truly interested in it. Most businesses or professions can become fantastically interesting when approached with the right attitude and all professions and businesses reward knowledgeable and innovative people.

If you are not currently involved in a profession or business, choose one and work toward it. Keep in mind that you can always change your goals, but have a clear idea of where you are going or at least be in the process of formulating goals. Simply by doing this, the probability of encountering more opportunities is increased.

23. We have the responsibility of initiating cause in our lives. For every effect there is a cause, whether that effect is positive or negative. We need not worry about success (effects) if we live each day congruently with our goals. There is little need to worry about the future if you do what you are supposed to do today. The future is for the most part determined by what we have done in the past and what we do today. Luck/fate is usually a minor variable in success. The most important variables are positive attitudes, planning, direction and hard work.

24. Be concerned about the natural world, the food you eat, and the air you breathe because if these deteriorate, so will you. Keep in mind that we are part of the natural world, and its laws apply to us. One need only view the statistics concerning unwanted pregnancies, deaths from alcohol or drug abuse, etc., to conclude that many do not believe that the laws of probability and physiology apply to them.

P: The Initial Interview, History Taking, and Selection of Clients

Long-standing, maladaptive behavior and the associated mediating and reinforcing emotional states can sometimes be changed in a few days or weeks. However, before I accept clients for my short-term therapy approach, they must meet certain requirements. An important requirement is the desire on at least a conscious level to be rid of the maladaptive behavior. Clients must perceive that their behavior is preventing them from experiencing a fulfilling life. This implies that clients have goals or are involved in some activities which they find enjoyable. Clients must have acquired some living skills that will enable them to experience positive reinforcement when they abandon their maladaptive behavior.

Clients who are not motivated can be made to understand that their maladaptive behavior is interfering with having an enjoyable life, but it may take considerable time to initiate and shape the necessary adaptive coping behavior. The important work and social skills which add meaning and dimension to life and underly the motivation to stay improved are not quickly learned, but of course can be over a longer period of time (3--6 months). Clients I rejected were those who exhibited signs of a chronic brain syndrome that resulted from starvation, or those in immediate medical jeopardy. Again, I will not treat are those diagnosed with borderline personality disorder. I attempted treating two such cases and, after a short time, abandoned treatment. Borderline personalities do not work well with my short-term approach.

Family therapy may be the best treatment for preadolescent and adolescent A-Bs. However, the validity of my approach has not been tested on this particular population. The vast majority, if not all of my clients, have been chronic adult cases who have been in psychological or psychiatric treatment for extended periods of time but were not progressing. Possibly a combination of family therapy and individual hypnobehavioral therapy would prove most beneficial for clients who are still in the home, but to my knowledge, these treatment approaches have not been combined.

Family conflicts such as anger toward parents may be important exacerbating variables causing and maintaining A-B even in clients who are not living at home. Therefore, these conflicts may need to be resolved. This may be particularly important for young adults who are in the transitional stage of leaving the protection of the family and becoming independent. When physical distance presents a problem and the parents are willing, much can be accomplished in resolving these conflicts through telephone conversations between the therapist, parent(s), and client. However, many parents are uncooperative. In many of my cases the families had avoided family therapy even at the onset of the disorder, indicating that the family therapy approach would not have been given a chance. This is a major flaw in the family therapy model.

I have treated mainly chronic bulimics who tend to be older, more impulsive, psychopathic, suicidal, and outgoing, and have a poorer prognosis than anorexics (Garfinkel, Moldofsky, & Garner, 1977, 1980; Morgan, 1975; Halmi & Brodland, 1973; Beaumont & Smart, 1976; Casper, Eckert, Halmi, Goldberg, & Davis, 1980). I had little contact with chronic A-B clients' families as many were adults living away from home.

All clients, especially those from out of state, are interviewed initially over the telephone. When a client is willing to call me and discuss her problems or is willing to send a description of her present condition and past history, it usually indicates that she has enough conscious motivation for treatment. Many chronic adult A-Bs are motivated simply because they are frightened concerning the physiological consequences of their behavior. In addition, many chronic clients are simply more mature than adolescent clients and thus better able to realistically view their problems.

For instance, I encountered four A-Bs classified as chronic and untreatable because they had exhibited A-B behavior for over ten years. All four were dangerously low in weight because of the chronic binge-eating and vomiting. Three were happily married, had children, and were deriving little secondary gain from their problem. The unmarried client was extremely emaciated, had been repeatedly hospitalized in the past, and had been classified as uncooperative and resistant. However, she was assertive and involved in a career which she enjoyed, and felt that the A-B was getting in her way. Three of the four were released after only two days of treatment. The fourth lived nearby and underwent 25 hours of therapy over a period of one month. Of the four, three were 95% free of the maladaptive behavior at an 18-month follow up, and the fourth rated herself as 80% improved on a one year follow-up. Since 1982, I have treated many more similar cases with similar results.

Variables such as a good marriage, job interest, and assertiveness are not a foolproof means of determining whether or not a client is unconsciously planning to sabotage the therapy. Most chronic A-Bs derive at least some secondary gain from their behavior and are often unaware of their motives. Also, clients who are able to form a good lifestyle and enter into meaningful relationships with others may simply be more resourceful and better able to alter their behavior. Increased motivation is probably only part of the answer.

It needs to be restated that clients who are so obsessive, confused, psychotic, or starved that they will not or cannot understand the therapy plan are poor candidates for my treatment and, as stated previously, I do not accept them. Also, some clients derive secondary gain from proving they can defeat all therapies. Of course this attitude must be confronted and altered from the onset. Clients with a borderline personality disorder diagnosis fall into this category.

Chronic clients with impoverished lifestyles and strong secondary gains can be successfully treated with a long-term hypnobehavioral program. The techniques presented form the basis of all successful psychotherapy, short or long term. Motivation to improve can be initiated, shaped, and reinforced, but this takes a longer period of time. Again, by long term I mean 3--6 months, not 1--4 years.

Prior to treatment, all clients received an audiotape describing in detail the treatment methods. The tape explained abreactive extinction, FRP, hypnosis, and separation therapy. Clients often remark after listening to the tape that, "Finally, someone really understands what I'm feeling and what is happening to me." This helps to establish rapport before therapy begins. Some clients experienced fear when they realize that they must encounter their anxieties, but to my knowledge this has caused few to avoid therapy. More often clients state that their expectation of changing is strengthened after understanding how the therapy works. Here is a transcript of what I typically say to introduce clients to therapy:

> "Before arriving, it is important that you understand and give some thought to what you will be doing in therapy before you arrive. First, let me explain a little about anorexia nervosa and bulimia. Although I am sure you are well aware of many of the details, I would like to review them with you. I understand that you were first troubled with anorexia and then became bulimic -- this is a typical pattern. As with most anorexics and bulimics, I assume that you obsessively fear gaining weight. In other words, you have accepted the suggestion or the idea that being thin will solve most of your problems.
>
> You most likely also feel that gaining weight is just about the worst thing that could happen to you, so that even the thought of gaining causes extreme fear. These feelings are very powerful, and mediate -- meaning, they cause -- you to binge and vomit and to avoid situations that are enjoyable. These emotions that drive you into losing weight also produce the sensory distortions of feelings of bloating in the stomach, thighs, face, and various other parts of your body which you may experience after you have eaten. These reactions are called conversion reactions, which means the anxiety has been converted into these feelings in order to keep you from eating so you will not gain weight. I suppose at times you have felt this bloating immediately after you've eaten. If so, this is your unconscious mind's way of preventing you from gaining weight. The unconscious mind has converted the negative emotions into these feelings that in turn keep you from eating too much or keep you from eating so you can continue to lose weight or

maintain what you consider to be a 'magical weight.'

However, at some time in the past you probably began to realize that unless you ate you would suffer serious complications; you then attempted to eat. When you ate, you experienced the extreme fears elicited by the fear of weight gain, so you then vomited in order to reduce those fears. The act of vomiting and eliminating food caused you to temporarily relax because you felt that this prevented you from gaining weight. But, because you tend to judge yourself very severely, you felt guilty about vomiting. The guilt was a strong negative feeling which you reduced by binge-eating again.

Food has been associated with love in most of our backgrounds, so the act of eating is relaxing and takes our minds off our problems. It is therefore understandable that when you are upset or feeling guilty you would want to resort to eating to relax. This is also what you most probably did as a child. Food also means calories, however, and calories mean getting fat. Since fat is what you fear becoming most, food now elicits strong fear through association as well as relaxation. This fear of gaining weight results from your acceptance of the idea that everyone will reject you and that you should hate yourself if you become overweight.

The overwhelming urge that you feel before you binge is a combination of the guilt and fear of gaining weight and fear of rejection. What you are doing through the binge-eating and vomiting behavior is avoiding negative feelings, but it doesn't work very well because these negative feelings always come back. In other words, you are only temporarily reducing these feelings by the act of binge-eating and vomiting. The avoidance of these feelings or partial avoidance of them allows them to continue to exist, so that you keep experiencing them. This explains why a person can get into repeated binge-eating and vomiting.

You should now be able to better understand how a vicious behavioral cycle is formed when negative feelings hit you that you do not understand. The stimuli and emotions which trigger these feelings are often from the unconscious mind. Whether you are aware of these stimuli and emotions or not, they do occur, and cause the urge to binge on food in particular as it has been paired in your past with love and affection. When you felt unloved, unworthy, tense, uptight, or whatever, one of the ways you reduced these feelings a child was to eat; because of further conditioning, food later caused strong emotions which in turn caused you to vomit in order to get rid of them. Because the act of vomiting itself causes negative feelings and guilt, you then felt an urge to again eat in order to relax. The more a person repeats a particular behavior or habit, the more automatic the habit becomes.

The extreme anxiety that you have been feeling has likely caused you to enter what are termed dissociative or hypnotic-like states. you've probably noticed when you are binge-eating and vomiting that time becomes distorted -- hours may go by like five minutes. This is one of the best indicators that a person is entering dissociative or trance states. I want you to understand that this does not mean you are psychotic or insane. People who have a good imagination and who are good hypnotic subjects can enter trance or hypnotic states easily. The similarity of the dissociative states that you have been experiencing and hypnotic states explains why we use hypnosis with anorexic and bulimic clients.

Through hypnosis, we can penetrate those trance-states and make you aware of what you are doing, as this is when much of your maladaptive behavior occurs. You may or may not be aware of it, but you probably regress to a younger stage of development during times of stress or during binge-eating. In other words, there is a child inside you which is angry, hurt, and/or frightened; when tension and anxiety increase, you automatically regress back into this more primitive mode of functioning. Keep in mind that most people do regress to a more primitive way of behaving when stressed.

I also assume that you have forced feelings and thoughts out of your conscious mind which you felt were undesirable, and that you judge yourself severely for having these thoughts. These thoughts and emotions or motives then become unconscious. Through hypnosis you will be able to understand your unconscious motives that are mediating your problem. You will also be able to contact the part of the personality which takes over during the binge-eating and vomiting or the starvation phase.

Think about it -- do you sometimes feel that you are totally out of control and that some other part of you is taking over, or that you feel as if you are observing your own behavior like an outside spectator? Remember, these feelings are just indicators that you do experience dissociative states. Keep in mind that I am not saying that you have a multiple personality or anything like that. Dissociation and time distortions are experienced by many creative people.

Understanding our unconscious motives is seldom enough to eliminate a strong habit pattern. The emotions or negative feelings which drive you into binge-eating and/or starving and comprise the urge to binge are learned or conditioned responses. By virtue of the fact that they are learned, they can be extinguished -- in other words, unlearned. It is important for you to understand that the active avoidance of the negative feelings which drive you into starving or binge-eating and vomiting prevent these emotions from being extinguished. You will therefore have to confront your feelings and experience them instead of allowing yourself to avoid them. By doing this, they will extinguish.

You will learn that they are only feelings and, although extremely unpleasant, they will not hurt you and you will not lose your mind by experiencing them. We will use hypnosis to regress you back to these feelings in order to bring them into consciousness so you can feel them and extinguish them. In other words, you will gain control over these feelings. This procedure is called 'abreactive extinction'.

You will also be undergoing what we call 'in-vivo extinction,' which is also termed 'response prevention and flooding.' This means that extinction will be done with food in real environments. We will be triggering some of the negative emotions by having you sample certain high calorie foods, and by having you sit through these emotions in my presence, they will extinguish. The goal is to not make you fat, but to free you from the negative emotions and maladaptive eating behavior so that you will be able to maintain a healthy, attractive weight. The motivation for maintaining the healthy weight will be self-love and improved health rather than fear. You are going to eliminate fear as the motivator for maintaining a good weight, which means you'll be free of those tensions and feelings that are preventing you from having the kind of lifestyle you desire. You must keep in mind that although these feelings are strong and facing them will take courage, they will never hurt you or cause you to lose your mind -- this, I can guarantee. You will feel a sense of relief and freedom after extinguishing these feelings.

In our unconscious minds there are often plans or motives that we are unaware of, and you may be deriving secondary gains from your behavior. By this, I mean that you may be unconsciously or consciously obtaining some gratification from your binge-eating and vomiting or starving. For example, you may be punishing someone else as well as yourself, or gaining attention from the behavior. The unconscious purposes or motives which cause you to maintain your maladaptive behavior will be be made conscious through hypnosis. You will understand these motives and become aware of your own feelings and thoughts so you can gain control of your life. You will be able to view yourself in a loving and more accepting way.

Through hypnosis and trances you will contact the part(s) of your personality that you are unaware of which is using the behavior as a weapon against yourself or someone else, or that is obtaining reinforcement in some other way for your behavior. You may discover that a part of you fixated at a young age binges, and an older part, usually of adolescent age who fears weight gain more than anything else, then takes control and causes the vomiting. Everyone's mind works in different ways. Some clients can readily visualize the fragmented parts whereas others cannot. For clients who cannot visualize the other part(s), flooding and response prevention procedures along with abreactive extinction may be all that is necessary for lasting changes to occur.

During our sessions, hypnotic recordings will be made for you to listen to at home in order to help you consolidate the gains that you make during our sessions. When you return home after your first three days of therapy, you will listen to at least one once per day.

Within a few days after receiving this, I'd like you to make an audio recording for me explaining your history, your problems, and your family in detail. Now I don't expect you to have all the material read and understood between the time you listen to this and when you make an audio recording for me. Take your time and describe your history and background, such as past treatments, what insights you've gained, and what you've felt was beneficial to you and caused any changes in your eating behavior. Describe your current behavior pattern in detail, plus what events or situations you've noticed that cause increases or decreases in your eating behavior. Also include information about your family history or whatever else you feel is relevant to your problem. Additional important information could include your relationships with boyfriends, friends and family, and your religious upbringing and current religious beliefs. Doing this will give me a head start on understanding and working with you.

We will be communicating often once we begin therapy, and I'm looking forward to seeing you."

It is important during the first day of treatment and in the introductory information that I convey to each client that she is accepted and that her behavior does not make her a 'freak' or a horrible person. Realistically there are worse behaviors that a person could be engaged in. This is extremely important, as many clients fear that even therapists are going to look down on them. Honest communication from clients is necessary in order to determine what specific aspects of the client's cognitive and overt behavior need to be modified. Unless the client believes that the therapist accepts her and is not morally judging her, she will not freely communicate her feelings.

The client's general history is reviewed during the initial interview and relevant situations are discussed; the client is also asked if she has any questions about therapy. The extent of the history taking will depend upon how extensively her history was presented on her audio recording sent prior to the first session. The client is asked to describe again in detail what she is currently feeling and doing, and a detailed behavioral history is taken. Questions as to when her starving or binge-eating and vomiting began and what was happening around that time may be important. 'Why' questions have little use at this point because, as with most chronic A-B clients, she is most likely unaware of the reasons for her behavior and will distort what reasons she may be aware of in order to avoid honestly confronting herself. Chronic clients seldom consciously recognize what secondary gains are being satisfied by their behavior. As previously stated, the behavior may be being used to gain attention, avoid adult responsibility, to punish someone else or themselves, or to dominate the home environment. These issues should be approached only after rapport has been established. I also ask what they have learned from prior therapies.

In addition, the client should be asked what happened before the onset of the maladaptive eating behavior and when it first began. This often provides situations to regress the client to in order to elicit the mediating emotions, and may also uncover the client's unconscious motives for holding onto her behavior. A complete history of the client's drug use, laxative abuse, illnesses, attitudes toward sex, religious upbringing, and current beliefs should be included.

Special attention is paid to variables that coincided with changes in her behavior. Clients, especially females, should be asked if their father was a heavy drinker. If the answer to this question is 'yes,' then it should be asked how their parents' sex life was. If she answers that it was not good, the therapist should ask how she knows that fact, and consider the hypothesis that she may have been sexually molested by the father. In my experience, few fathers sexually molest their daughters unless they are intoxicated, although few commit this crime even then. However, ignorant MHPs have blown this way out of proportion and subsequently ruined many people's lives. It is difficult or impossible to prove that a client was molested as a child and has not instead fantasized it as a result of suggestions given by popular media or prior MHPs. I have encountered only a few A-Bs who have been reported being molested.

Information gleaned from hypnotic regressions is often distorted, but it does not invalidate hypnosis as a therapeutic tool. The memories encountered are real to the client, can mediate their maladaptive behavior, and should be worked with as the information gained through hypnotic regression taps into what is or was in people's minds at the time. This is reality to them and can often guide their behavior and thinking.

Again, very few of my clients alleged sexual molestation or experienced forgotten molestations through hypnotic regression. I have had only two clients who did remember molestations by close relatives, but a year later, one totally recanted her story. I had not believed she had been molested because, as therapy progressed, her stories had become more and more fantastic; as stated, human memory has been proven to be very unreliable. I therefore focused more on in-vivo extinction, as I do with all my clients.

The theoretical basis of hypnobehavioral therapy and the etiology of A-B is explained during the initial two hours of therapy. It is best to reinforce the concepts by citing examples from past cases and from the client's own behavior. I try to keep the descriptions and theoretical explanations congruent with the client's level of understanding.

I explain their conversion reactions and how they maintain maladaptive behavior, with the client's specific conversion feelings used as examples. Trances and dissociative states are explained in detail. The client must understand that entering trances blunts emotions and, by dissociating the emotions, a fragmented part of the personality may stay fixated at a particular stage of development and take over during the maladaptive behavior. Explaining how behavior can be broken down into drastic changes between mental states that coincide to different behavior patterns is often helpful. With A-Bs, one state coincides with starving, whereas one or more other states correlate to binge-eating and vomiting. I explain that the purpose of therapy is to construct a normal state somewhere between the two maladaptive states. The origins of dissociation and the separation anxiety which subserves their problem behaviors are explained. The client must understand that whether her parents actually rejected her or peers convinced her that she has to be thin to be acceptable, or whether she incorrectly concluded that this happened, the strong negative feelings are the same and these feelings mediate maladaptive behavior. Clients must also understand that they must become aware of the secondary gains that reinforce their maladaptive behavior and cause resistance to change. Examples of how they may be using their behavior as a weapon against themselves or others is given.

Following the description of hypnosis and the behavioral concepts, a client's religious beliefs should be understood. If she is religious in the orthodox sense, a Christian approach is taken concerning self-love and so forth. For non-Christians, the philosophy of self-love, tolerance for themselves and their neighbors, and abandoning the 'perfectionist's script' is shown to be essential for having a good life. If the client is of a non-Christian faith, these same principles can usually be gleaned from their religious scripts. The client's religious beliefs should be built upon whenever possible, and only directly challenged when they are unadaptive and incongruent with the basic Christian teachings concerning self-love and guilt.

Again, MHPs should connect with educated, responsible religious people who can give counseling in these areas. I have never had a problem with any educated religious counselor except for a Mormon Bishop. I had contacted this Bishop to clarify a Mormon doctrine. I argued with the Bishop because his concept of sin and God's forgiveness was incongruent with the Christian view. The phone consultation, which the client was privy to, did prove helpful.

FRP and hypnotic abreaction are again explained in detail even though the client has supposedly read a description of them. This is important because these procedures are frightening and, since chronic clients resort to extreme dissociation and trance-logic in response to anxiety, they may conveniently 'forget' what they have read or simply play stupid. Additionally, many clients did not read the information provided and had only listened to my recorded explanation prior to coming in.

For example, a 19-year-old chronic bulimic travelled from out of state for therapy, but then refused to participate in FRP. This was despite the fact that she had listened to the introductory explanation tape and read the required material before coming in. Her reason for refusal was that she had asked her mother if those procedures were really going to be used; her mother had replied that she didn't think so, but didn't really know. The client used her mother's answer as proof they were not going to be used and as an excuse to avoid the FRP. Hypnosis was attempted with little success, and she was discharged after one day of therapy as a result of her uncooperative attitude. On a year's follow-up, she had not made any progress elsewhere.

However, I encountered a few clients who did not seem to care about theories of treatment. Their attitudes, as stated by one, were, "I am tired of this problem so let's just get on with it." One client was in her middle fifties and on the verge of emaciation. She had experienced A-B behavior since she was 19 years old, and expressed this attitude. After three days of treatment, she was 95% better. Six months later she came in to visit me as she was traveling nearby. She was very happy with the gains she had made and told me that I did not charge enough, as she had spent so much on other therapies that had not helped.

Re-explaining the concepts many times may be needed with some because they have been told by previous insight-oriented therapists that hypnosis and behavioral methods do not work and are dehumanizing or dangerous. These negative expectations can be overcome to at least some degree by a thorough explanation of the hypnobehavioral concepts and by using past cases as examples. The regressive aspect of hypnosis and the basic behavioral laws should be explained in detail. The hypnotic and behavioral models for OCD should also be explained and illustrated by examples.

Dreams and their significance are also explained because as therapy progresses clients may experience periods of fragmented sleep and more dreams than usual. Admittedly, much of the following is speculative because the psychological significance of dreams is still not well understood. My interpretation of the clinical significance of dreams has been partially gleaned from Kappas', et al., observations as well as my own.

Post-hypnotic suggestions are given in order to trigger venting dreams toward morning. The client is told under hypnosis that morning dreams often vent resistances and emotions that may be impeding the client's progress. On the first day of therapy, the client is given a suggestion that she will awaken with a dream toward morning that will vent a conflict, fear, negative emotions, or whatever may be inhibiting her from obtaining her goals. She is also told that when she awakens with the dream she will write it down in detail and then, and only then, go back to sleep and awaken later refreshed. Dream suggestions are routinely employed by hypnoanalysts who use the dream material to interpret unconscious resistances. When therapy is at a standstill, a dream may be triggered which, with the aid of hypnotic and word association procedures, can reveal unconscious resistances that are sabotaging the therapy.

I have seen no real evidence that dream venting suggestions are that useful in actual therapy. However, it is important that clients understand that experiencing nightmares and/or increased fragmented sleep during which more dreams are remembered is a sign that they are getting better, as REM rebound is a sign of the brain healing. Rather than fearing nightmares, therefore, they need to realize it is a sign they are getting better.

I require clients to keep a log of what they eat and drink, which has proven very helpful. There are many foods and additives that exacerbate anxiety, etc. Many types of symptoms can be triggered by these substances. Hypoglycemia and the importance of good nutrition is also thoroughly explained on the first day and a hypoglycemic diet is recommended. The typical explanation and further discussion is contained in the next chapter.

Lastly, I explain that some of the hypnotic inductions will involve using an arm-raising technique involving the non-dominant hand as it is influenced by the non-dominant hemisphere of the brain. I tell clients that this is done because it is suspected that much of the unconscious mind (visual imagery, imaginative power, and non-verbal functioning) resides in the non-dominant hemisphere. This helps the client to better understand that the MHP is attempting to have her accept suggestions and reprogram her unconscious using her imagination and nonverbal skills. This is based upon much theoretical speculation, but it is the best explanation that I have at this time for the results stemming from this technique.

Q. Important Non-Psychological Variables

Since humans are biosocial creatures, it is reasonable that psychotherapeutic approaches based solely on a learning model may have limited utility in the treatment of organically based psychopathology. There is an interaction between environmental or learning variables and organic variables, including hereditary predispositions and the nutritional status of the brain. A person who is feeling physically well and who is maintaining an optimum nutritional intake and regularly exercises should be more energetic and more motivated to learn new coping skills, and as a result will experience better results in psychotherapy. Therefore, approaches which combine improving organic conditions that influence behavior along with learning adaptive behavior are most reasonable.

The influence of nutrition on behavior has been a subject of much controversy. Various vitamin deficiencies have been suggested as possible etiologic factors in psychiatric disorders (Milner, 1963: Schulman, Sterner & Price, 1973; Pauling, 1968; Watson, 1972; Hawkins & Pauling, 1973, Cheraskin & Ringsdorf, 1971). Some qualified writers claim cures for almost every disease and emotional problem by simply improving clients' diets or through the administration of megadoses of vitamins and minerals (Fredericks, 1976; Cheraskin & Ringsdorf, 1971; Cheraskin, Ringsdorf, & Brcher, 1974). The current evidence for some of these claims is meager and often methodologically inadequate. However, it is also premature to conclude that in at least some individuals even mild vitamin deficiencies could not be a significant etiological variable. Although it is rather rare that the addition of specific nutrients either in small or megadosages causes miraculous cures, it is also unlikely that an individual's overall nutritional intake has no effect on states of mind. For example, thiamine deficiency has been shown to reduce glucose absorption in the central nervous system by as much as 50 to 60 percent (Guyton, 1971), and experimental deprivation of thiamine can cause mental disturbances characterized by depression and anxiety (Pike & Brown, 1956). Optimum physical health obviously has a significant positive effect on people's emotional sense of well-being and level of energy, which will in turn affect the degree to which they are motivated to comply with behavioral assignments.

For the past 39 years I have recommended that people take a complete multiple vitamin supplement. For memory retention, I recommend DMAE, a precursor to acetylcholine. I also have seen other supplements in varying situations work well; a type of L-theanine in particular greatly reduces anxiety (often quickly) without any side effects. Some clients may be suffering from excessive vitamin and mineral intake due to the ongoing fad of megadosing of vitamins and minerals. It is reasonable to assume that if vitamin and mineral deficiencies can exacerbate or directly cause emotional problems, excessive amounts of some agents may also be deleterious to some people's central nervous system. Some who megadose themselves with vitamins and minerals may experience a variety of 'mental' symptoms such as dizziness, disorientation, and insomnia.

I encountered three clients within a period of one year who reported the above symptoms as a result of the ingestion of a vitamin-mineral supplement containing zinc and several B vitamins. One client in particular experienced major disorientation and became bedridden as a result. When the supplement was eliminated, the symptoms remitted in all three cases. The long-term effects of megadosages is poorly understood, and moderation is the best advice that can be given when consultation with an expert is not possible.

Some medical authorities dispute the claim that poor nutrition may play a significant role in the etiology of diseases affecting westernized cultures and have stated that the average middle-class North American is well fed and receives adequate nutrition from his daily diet. However, being 'well-fed' may not be equivalent to being well nourished. For example, it has been found that a large number of Canadians suffer from mild to moderate nutritional deficiencies (Sabry, Campbell & Forbes, 1974). The environment may be polluted with a variety of toxins which can inhibit the effective synthesis of protein and other nutrients as well. There is no doubt that certain heavy metals are etiologic factors in some central nervous disorders. These possibilities should be ruled out by a physician through blood tests, hair analysis, and whatever other diagnostic tools are available. In my opinion, all clients ideally should be screened by a medical doctor who specializes in nutrition and be treated with supplements as necessary.

MHPs often encounter clients who are adhering to potentially dangerous extreme dietary regimes in order to lose weight. If all or nearly all carbohydrates are eliminated from a person's diet, ketosis will occur. This occurs because the body has utilized fat as energy and produced an accumulation of ketones in the blood and interstitial fluids. The result is the development of acidosis, which causes depression of the central nervous system. However, a low carbohydrate diet if done within reason can be safe under a doctor's supervision. The counterpart of acidosis is alkalosis and can result from the excessive intake of alkaline substances, which may cause traumatic hypersensitivity of the nervous system. These two conditions can drastically affect the client's emotional stability and should be remedied through proper medical and dietary intervention.

Occasionally I encountered a client who was extremely obese and remained so despite the fact that her food log indicated a low caloric intake. There are at least three possible explanations for this: 1) the client may not have accurately kept the log in regard to the amounts and/or caloric values of the food eaten; 2) she may have entered dissociative states while binge-eating and not consciously remember doing so; or 3) she could have a metabolic problem or be taking medications which prevent weight loss. The second possibility is less common than the first, although I have encountered clients who regressed to a childlike state during which they binged without any conscious recollection of doing so. These clients are similar to bulimics and respond well to the hypnobehavioral approach.

Hypothyroidism may explain why some clients who maintain a low caloric intake do not lose weight. The symptoms of hypothyroidism include fatigue, depression, and headaches (Barnes & Galton, 1975). For 39 years I have referred people to physicians to check their thyroid. A significant number of people were hypothyroid and, after receiving medical treatment, felt much better and were able to lose weight.

It is generally accepted that most adult Americans ingest more protein than is really needed. This generalization may lead one to believe that the behavioral manifestations of protein deficiencies are relatively uncommon. However, large quantities of protein can be consumed by some individuals but they may still be deficient in net protein utilization. This deficit may be due to the fact that they are eliminating amino acids which must be present in the appropriate ratios for the optimum absorption and utilization of protein by the body. Interestingly, physiological and emotional stress have been shown to increase the need for protein as well as other nutrients. Physical activity is also important as vigorous exercise is almost always followed by a strong sense of well-being and can be a potent antidepressant. One study showed that a group of depressed students who were put on a six-month jogging program did better than a group who underwent individual and group therapy.

For chronically stressed clients, physical activity is an efficient means of normalizing the biochemical aspects of stress. It is reasonable that the 'fight or flight' emotional state should be followed by some type of vigorous physical activity and that activity should bring the system back to homeostasis. We evolved to exist in a hunting-gathering type of society. When humans experienced a fear stimulus, it was usually followed by extreme physical activity such as running or fighting.

When one considers the amount of junk food that many Americans consume and the alarming rise in diabetes, psychotherapists should explain to clients the dangers of ingesting large quantities of refined carbohydrates (in particular white flour and all sugar), caffeine, and alcohol. It should be pointed out that overconsumption of these foods may cause hypoglycemia (low blood sugar) which in turn can exacerbate psychological disorders. During the FPR exercise, ingesting the required simple carbohydrate foods such as doughnuts and cookies often causes the client to become sleepy and mentally foggy. It is during this foggy state that physiological cravings for simple carbohydrates occur because of the blood sugar drops. While some medical authorities in the past felt that there is no such disorder as hypoglycemia or that it was uncommon, it has been my observation, along with many physicians who work daily with emotionally troubled people, that it can be a major exacerbating variable in depression and anxiety problems.

In hypoglycemics the ingestion of refined carbohydrates, caffeine, or alcohol chronically stresses the pancreas, which oversecretes insulin. One of the functions of insulin is to convert excess blood sugar to body fat. When hyperinsulinism occurs, the blood sugar level may fall below the level necessary for normal central nervous functioning. The brain accounts for less than 3% of the total body weight, but it consumes as much as 25% of the available blood glucose. Therefore it is highly sensitive to blood sugar fluctuations and dependent upon adequate glucose reserves. When the blood sugar level falls because of either fasting or the over-ingestion of refined carbohydrates, alcohol, or caffeine, hypoglycemics may experience a ranges of psychological symptoms, including fatigue, dizziness, headaches, tremors, mental confusion, lethargy, and depression (Fredericks, 1976; Cheraskin, Ringsdorf & Brecher, 1974).

The six-hour Glucose Tolerance Test (GTT) can be administered by a physician to diagnose the "true" hypoglycemic whose symptoms can be alleviated through dietary adjustments. Unfortunately, the GTT may not tell the whole picture. Evidence indicates that blood sugar level drops may not correlate to the symptoms and subjective distress reported by clients. Apparently many of the symptoms can be due to hyperinsulinism alone as a surge in insulin causes an uptake of fluid by the brain tissue. This factor is not monitored by the six-hour GTT as it only monitors the levels of blood sugar. Additionally, clients may be allergic to certain foods or chemical additives which can cause blood sugar fluctuations also. Therefore, the ingestion of glucose alone may not cause the same blood sugar fluctuations as would the ingestion of certain food(s). It is often simpler to try the hypoglycemic diet for a month to see what benefits are derived.

Emotional stress may also cause alterations in blood sugar levels. It is possible that with some clients there is an interaction between emotional stress, poor diet, and allergies, so that the symptoms appear only when all are present. This is not to say that the GTT is worthless. The rate of the fall of individuals' blood sugar levels may be more important than the absolute level, and a knowledgeable physician can interpret other aspects of the blood-sugar curve. Some physicians now give tests measuring mental ability periodically throughout the test to more precisely test the effects of hyperinsulinism on clients' mental functioning. Again, I refer all willing clients to a doctor who understands these patterns.

How a random blood sugar test may be misleading in assessing the influence of hypoglycemia and other reactions to food on a client's emotional and subjective state was demonstrated by a teen-aged male client. After having been referred to numerous medical specialists with unsuccessful results, he was referred to me. The young man experienced catatonic-like states during which he had difficulty moving, speaking, or getting out of bed. These attacks often incapacitated him for as long as a week. The referring neurologist had checked for low blood sugar during an attack, but found the blood sugar level was normal.

I took an extensive history which revealed that the young man had experienced his first attack following a severe migraine headache. The migraine had coincided with his ingestion of large amounts of Chinese food at the restaurant where he was employed. In some people monosodium glutamate

(MSG), common in many foods and particularly in Chinese food, can trigger migraines. A survey of his eating habits revealed that he consumed at least two liters of soda per day along with many candy bars. He was overweight, and at times craved food during his bizarre attacks.

I advised him to eat according to the hypoglycemic diet, which eliminates all caffeine, sugars, refined foods, and MSG. The attacks immediately subsided for three months. However, he then began to violate the diet and soon experienced another attack. When he once again maintained the dietary restrictions, his attacks subsided and stayed in remission. It should be kept in mind that this client had manifested no psychopathology during the initial interview or when he was not experiencing the attacks. However, when an attack did occur, he showed irritability, extreme lethargy, mental confusion, and pseudoseizures were suspected.

I cannot state positively that the placebo effect was not operating in this particular case. It is possible the suggestion that these foods could have caused his problems was enough for him to unconsciously give up his bizarre symptoms. However, this client had not proved hypersuggestible nor had he obtained any relief from previous treatments. It should be noted that occasionally clients appear to be poor hypnotic subjects until defenses diminish and their true suggestibility emerges. In any case, placebo effects usually diminish over a six-month period. At last contact with this client, over a year had passed with only mild recurrences of his problem when he violated the diet.

Therefore, when hypoglycemia is suspected, clients should be referred to a sympathetic physician who is knowledgeable about glucose absorption disorders and who employs the proper diagnostic tools. If there are no appropriate physicians available, MHPs should feel secure in recommending a simple hypoglycemic diet to clients who are not already on a prescribed diet. There are no dangers to eliminating sugar, white flour and other refined carbohydrates, alcohol and caffeine from one's diet. The diet below is the one I usually recommend after consulting with the client's physician.

After awakening in the morning, one glass of unsweetened fruit juice should be drunk, followed by one of the following high protein foods: cheese, eggs, meat, fish, fowl, seafood, or any complete protein product, or whole grain cereal with milk if possible. Clients should snack on a high protein food or unrefined (complex) carbohydrates whenever they experience hunger. They should not allow themselves to become overly hungry. Complex carbohydrates include whole grain products and all unprocessed vegetables. Processed prepackaged foods should be avoided. Meals should be small and clients must eat small amounts every several hours in order to keep their blood sugar levels stable. A green vegetable should be included along with high protein food during meals and, if weight gain is not a problem, no restrictions are given as to ingesting vegetables such as potatoes, corn, squash, etc., as all are unprocessed.

I also suggest that a large glass of skim milk, preferably warm, be taken before going to bed as it contains L-Tryptophan. L-Tryptophan helps people to sleep and the intake of protein before bed may help the blood sugar remain in a normal range throughout the night. (Alternatively the supplement L-Tryptophan or 5 HTP may be taken.) This allows individuals to awaken more easily in the morning with increased energy and may decrease fragmented sleep and night binge-eating in bulimics. Some clients have stated that they awaken anywhere from 30 minutes to three hours after falling asleep as a result of hunger. However, when they had a high-protein snack before going to bed, many reported that they awakened less often.

If clients insist on occasionally having an alcoholic drink, they are instructed to avoid beer, sweet mixed drinks, liqueurs, cordials, and sweet wines because of the high sugar content. Alcohol itself causes blood sugar drops and, when taken before going to bed, can increase awakenings with urges to binge and cause fragmented sleep. It should not be used as a sleep aid.

Some authorities feel that fructose as ingested as fruit (not high fructose syrup) may not elicit the hyperinsulin reaction due to the fiber. It is my opinion that this may or may not be true, and should be experimented with carefully. Clients are told they can eat a few pieces of fruit a day, but to not overdo it. Artificial sweeteners are also forbidden because of the possibility of eliciting conditioned hyperinsulin reactions, although these reactions should extinguish over time. The possibility of artificial sweeteners causing their own set of problems because they are artificial still remains however.

Almost everyone who stays on this diet for a few weeks report that they feel better and have more energy. Clients must be warned, however, that extreme fatigue and cravings for sugar, caffeine, or refined carbohydrates may occur during the first three days. Most have found that even when this occurs, they are less irritable, anxious or depressed, and feel emotionally more stable.

Hypoglycemic episodes can cause increased sympathetic activity, which results in physiological sensations that mimic anxiety attacks. These sensations may have provided the US (unconditioned stimuli) for the negative feelings that had become associated with the stimulus conditions which presently trigger anxiety. I have traced some phobic reactions to what appeared to be hypoglycemic attacks. Negative mediating emotions, whether they occur as a result of conditioning, hypoglycemia, or the fear caused by the physical sensations of sympathetic overactivity, are still negative mediating emotions which clients attempt to avoid. Clients who are unaware that they are experiencing hypoglycemic attacks often label them as anxiety attacks, precursory sensations of a heart attack, or feel as if they are losing their mind. This causes more fear and sympathetic arousal, thus mediating more avoidance behavior. With some clients, the simple explanation of the "how's" and "why's" of the hypoglycemic attacks has caused their phobias and anxiety attacks to significantly diminish. Occasionally, simply recommending the hypoglycemic diet without suggesting to the client that it would eliminate their anxiety or depression has resulted in a complete remission of these problems. This is of course rare, and the effects of the clients' expectations (placebo effect) were not rigidly controlled for.

Hypoglycemic reactions being mediated through the autonomic nervous system and endocrine system can themselves come under stimulus control. Autonomic and endocrine responses can easily become CRs. Therefore, if people have hypoglycemic attacks in a particular situation, they have an increased probability of experiencing that same hypoglycemic attack in a similar situation. Conditioned hyperinsulinism explains why some clients experience strong urges (blood sugar drops) for refined carbohydrates when they simply see, smell, or imagine certain foods. These cravings for junk food, etc., can be extinguished by repeatedly presenting clients with the food or image of the food and by not allowing them to eat it. The extinction of the urges follows a typical extinction curve. The urges elicited by alcohol and tobacco stimuli may also involve hyperinsulin reactions, and, as stated previously, can also be extinguished by imagery and in-vivo extinction.

Hypoglycemia may be an important variable in binge-eating and may explain why many obese people grow hungrier as they become fatter. The more junk food they eat, the more they experience hyperinsulinism, and the hungrier they become. Weight is gained during this process because the insulin is converting the excess blood sugar to fat. Most obese clients respond well to the therapy techniques described in this book for the treatment of bulimia. Obese clients who experience overwhelming urges to binge-eat are simply bulimics who have not learned to vomit or have not acquired a strong fear of weight gain. The learning history of chronic bulimics and obese clients are often very similar. As stated previously, obesity may be maintained by secondary gain and unconscious motives such as the avoidance of intimacy with the opposite sex.

Most maladaptive behavior, in particular dissociative states, is exacerbated by hypoglycemic attacks. I have encountered numerous A-Bs and obese clients who awaken late at night and binge. Most probably some of these clients are experiencing blood sugar drops and, therefore, real hunger. The feelings of hunger are the stimuli, which elicit the fear of weight gain, which in turn initiates the binge-eating and/or vomiting cycle. The hypoglycemia-induced urges along with the emotionally-based urges can be extinguished through extinction methods. This often results in a reduction of the nighttime binge-eating. The reader is referred to Case Study One for a demonstration of how the extinction is accomplished.

In conclusion, MHPs should establish a relationship with a physician(s) who is knowledgeable in the aforementioned areas. There are many good diagnostic tools such as blood tests that can help elucidate vitamin deficiencies and hormonal problems that cause people to feel unwell, depressed, and fatigued. Optimizing these variables can be very important.

III: Case Studies

The purpose of this section is to present examples of the hypnobehavioral techniques so that readers can judge for themselves what is taking place. It is one thing to read a theory of psychotherapy, and quite a different experience to read the actual dialogues or view the practitioner in action. As little editing has been done, the dialogues are long. I felt that it was better to be overly inclusive than incomplete. The reader should realize that therapy is often tedious; as clients often do not volunteer information, it must be drawn out of them. Editing was done in order to make the dialogues cohesive, but such editing is indicated.

I realize that it is difficult or impossible for the reader to assess a client's nonverbal behavior from the dialogues, but it has been my experience that clients' nonverbal and verbal behavior are usually congruent during the procedures. Of course, if the client's voice indicates she is experiencing little emotion during abreactive extinction, she is obviously dissociating the feelings and preventing them from extinction, or simply faking the whole procedure This maneuver should then be confronted and not allowed. This was rarely encountered in the following cases.

1. Abreactive Extinction of Emotional and Hypoglycemic Triggers Occurring During Sleep

The following is a transcript of the second to last session of therapy with a 36-year-old female who had been binge-eating and vomiting throughout adulthood. Prior to entering therapy with me, she had binged and vomited on an average of twice per day. She was a somnambulist, and responded well to therapy.

Her self-esteem had greatly improved by this session, and she had been able to gain a few pounds which she had greatly needed. Her energy level had also greatly improved, and she was enjoying the praise from her co-workers and family concerning how much better she looked. At this point in therapy, her binges had decreased to two or less per week; these occurred only after falling asleep and the magnitude of the binge-eating and vomiting was greatly reduced.

A standard progressive relaxation induction was employed, after which the transcript begins:

K: ... Now, as you are relaxing and moving deeper into relaxation, I'm going to ask you to talk to me. I want you to talk to me without thinking, and when you awaken from the hypnotic state, you don't have to remember what you said if you choose not to. Just relaxing now ... and remember, you may forget what you said when I talk directly to the part who takes over during the binge-eating. In a moment, I'm going to count down from 5 to 0, and when I do, you are going to move into the same state of mind that occurs during those binges. In fact, the same part of the personality is going to take over... 5 ... 4 ... 3 ... 2 ... 1 ... 0 ... Now, when the part of the personality who takes over during the binges at night takes over, just let me know by nodding your head.

(Client nods.)

Why are you bingeing?

C: ... Hungry...

K: Why don't you eat just a small amount?
C: I have the urge to eat more.
K: Do you know why that urge is there?
C: It helps make me feel better.

K: What are the feelings that you are trying to get rid of? I know one is hunger ... do you awaken hungry and also tense?
C: I probably can't solve some problems.
K: Remember now, you don't have to remember anything when you awaken, so just talk to me automatically. How old is this part -- how old are you?
C: (no answer)
K: All right, you're tense when you awaken, right? And the food takes away the hunger and the tension?
C: Yes.
K: Go deeper now, deeper, sinking deeper ... I want you to just imagine that you're drifting back to Sunday night now, and you awaken, you're awakening and you're feeling certain feelings ... nod your head when you begin to feel the hunger.
C: ... I didn't really feel hungry...
K: What did you feel?
C: I felt I ate too much at dinner.
K: And when you awakened, how were you feeling?
C: I was feeling full and thirsty, so I went to get food.
K: Why did you go to get food when you were thirsty and already felt full?
C: I don't really know.
K: What did you eat Sunday night ... just look and tell me ...
C: The first time (I awakened), I didn't eat because we had had company and they were still there; I just got something to drink and went back to bed. I awakened at 6 0çlock and went to get some food although I still wasn't hungry, but my husband and a friend were still up. My husband was tired, so I said we ought to go to bed ... and his friend left. We went to sleep for an hour, and then I awakened again and binged; but I didn't eat very much, just a couple of pieces of meat and some vegetables.
K: What did you do then, vomit?
C: Yes. I felt guilty when I was vomiting though. It's harder and harder to do.
K: But the full feeling was uncomfortable?
C: Yes.
K: All right, now imagine that full feeling, feeling full. Now that you're imagining it, do your legs feel too fat and does your stomach feel full? Yes?
(Client nods her head.)
All right, let's go back to what triggered that full feeling. I'm going to count down from 5 to 0, and you're going to move back to what triggered that full feeling... 5 ... 4 ... 3 ... 2 ... 1 ... 0 ... What's coming into your mind now? What are you feeling?
C: That I ate too much, with all the food.
K: That you ate too much means what?
C: Fat.
K: Fat means what?
C: Ugly.

K: Ugly … So it was the fear of getting fat, wasn't it?

C: Yes.

K: When you awakened, you felt what first?

C: Full.

K: Full means what?

C: Fat.

K: And how do you get rid of fat feelings?

C: Vomit.

K: So that's why you didn't binge on very much food … You used just a little bit of food to vomit in order to get rid of those feelings, didn't you?

C: Yes.

K: Take a good look at this situation now. It wasn't so horrible, but there are better ways to maintain your weight. When you overeat, just eat less later. However, you overate because of, what? … Moving back again, and take a look … the food was good? -- or were you tense? -- or both?

C: Probably both. I don't know why I should have been tense to begin with, though, it was such a nice day.

K: But you just overate… What did you overeat?

C: Italian bread, some Italian vegetables, and put them on the grill.

K: Were they good?

C: Very good, and I had made an eggplant before that, which was also very good.

K: I get the feeling that the food was just very good and you just overate.

C: I'd had a fruit salad after that, which was very good too. There was just too much food.

K: All right, let's go back to Tuesday night now. I'm going to count down from 5 to 0, and when I reach zero, you're going to be awakening again as if it is Tuesday night, and we're going to be seeing what feelings or thought preceded that binge… 5 … 4 … 3 … 2 … 1 … 0 … What's coming into your mind now, what are your feelings?

C: Very full.

K: Feel the full feelings, imagine them, feel them. Now you're feeling your thighs swelling, stomach swelling … that's okay now… feel the fear … what's coming into your mind?

C: I don't feel so good.

K: Feeling fat?

C: Yes.

K: Feel yourself growing fatter and fatter now… that's it… feel the fear … do you want to hold onto my hand? OK, hold on… Feeling fatter and fatter now.

(The negative mediating emotions are being abreacted at this point.)

C: I feel as if I'm this big!

(Client indicates with her hands that the size of her stomach has greatly increased. This is a conversion reaction which she produced in order to prevent herself from overeating. Clients may unconsciously and consciously cause the intensity of the conversion reactions to increase in order to have a margin of safety since they believe it is much more desirable to be too thin than even a little overweight. Most anorexics and bulimics -- including this particular client -- overly rely upon the conversion feelings over time; this over-reliance often prevents the client from eating enough to stay alive.)

K: Now that you're big, how big do your legs feel?

C: (The client indicates with her hands how big she feels.) I've always felt that my legs were bigger than my stomach.

K: All right, feel them getting bigger ... becoming elephant thighs, hippopotamus thighs...

C: My calves, also, they hurt sometimes.

K: Keep feeling yourself getting fatter and fatter now, fatter and fatter. Look at you - fatter and fatter. Feeling fear?

(Client nods.)

Feel the fear of getting fat, fatter, and fatter.

(The conversion reaction is purposely intensified so the client can better identify the feelings and thought which elicit it.)

Let's find out what triggered that fat feeling: It occurred even before you ate, didn't it?

C: Yes.

K: It probably had nothing to do with eating. Now, again, I'm going to count back, and when I reach zero, you will be at the feelings, emotions, thoughts, time, place where this feeling began... 5 ... 4 ... 3 ... 2 ... 1 ... 0 ... Now, look ... what is it? What's coming into your mind?

(Clients are always told to "look" or "feel" and to not think. Critical thinking and analysis by clients helps them to dissociate the emotions, therefore impeding the extinction process.)

C: You've got me calmer now.

K: Give me your first thought...

C: My legs don't feel fat.

K: Well, you know because your mind brought it on, your mind can take it away. I wonder why you got fat even before you ate? ... Take a look now...

C: Seems like I was still so full from lunch. I really didn't feel like eating, but I knew I had to eat something.

K: And then you gave yourself the suggestion that you were going to get fat -- you expected to get fat because you overate. You're right in one way; when we do overeat consistently we will get fat. However, one time doesn't make us fat. What you're going to do from now on when you eat lunch is eat a smaller amount.

C: But I do eat a small amount of food.

K: And you still feel fat?

C: Yes. Well, it wasn't as bad today because I ate a couple of cucumber slices before I made my lunch, and then I made my lunch. I usually sit up at the counter on a stool, and I said, 'No, instead I'm going to take it to the table, and I'm going to sit there and relax as I eat,' and it helped a lot.

(The client was experiencing the results of posthypnotic suggestions to eat at the table and relax, which were on the audio recordings made for her by me at prior sessions. She had been instructed to listen to one or two of them every night before going to bed.)

K: It helps to relax when you eat, because tension brings a lot of this on, doesn't it?

C: Yes. I've also been craving bread, bread, bread, bread, like for breakfast. I think that the craving brings on a binge, so now I have a piece of toast with cheese on it, or a high protein gluten bread without sugars or anything with scrambled eggs or cheese for breakfast. It's very good.

K: Well, you've been doing very well, don't lose sight of that. You just relax, and go deeper into hypnosis and always keep that in mind, that you're doing very well. One binge doesn't make you bulimic, two doesn't either, so relax. It's interesting, isn't it, that the binges correlate to the family problems. You had mentioned to me that your family is still having a lot of problems and that your mother couldn't come over on Tuesday because of the problems, right?

(Client nods.)

Let's begin with Mom; what did you think about Mom not coming over?

C: I know why she couldn't come. She really wanted to, but Dad was having problems again.

(Her father was a chronic alcoholic, and the client had strong ambivalent feelings toward him.)

K: How did that make you feel? -- First thought?

C: I felt disappointed, a little hurt.

K: Feel the hurt, feel that hurt... Hurt is because of what -- a little bit of abandonment?

C: yes.

K: All right, you feel a loss of love. Mom couldn't come, so you felt a little bit unloved. Dad is acting up and you don't feel much love there... What are you beginning to feel now?

C: Tension.

K: Tension? When is the last time you felt this tension?

C: Just a few minutes ago.

K: When was that?

C: When we were talking about the binges.

K: It's pretty obvious now, isn't it? Let's talk about it, and begin to extinguish those feelings. You experienced the strongest feelings that a person can feel -- separation or loss of love and abandonment from a parent. These feelings were triggered when Mom couldn't come over, and you felt unloved and rejected. In addition your father's behavior causes you to feel unloved and at the same time angry. But you're afraid to express your anger towards any of your family because you fear that they will then reject you further. All these feelings build up, and then you reduced them previously by what?

C: By eating.

K: You don't need to do that anymore, because we are going to get rid of those feelings. You are loved! Your Mom had a problem with Dad, and Dad's got a problem, but that's not your problem. OK, I want you to think about it now -- think about Mom, think about Dad, think about the hurt. All of them hurt you because you are a very kind wonderful and compassionate woman, and you worry about them, don't you? --Also, you're human and you're angry with Dad?

C: Yes...

K: Now feel the tension growing...

C: Mom said that she was going to have to go down there and decide one way or another what she was going to do with Dad. She's got to be a strong person or she couldn't put up with what she has for so long, but I hate to see them separated because I know Daddy would just get worse.

K: So your ultimate fear is that Daddy is going to end up dying from being a chronic alcoholic?

C: Yes.

K: Feel the fear, it's a natural thing, feel the fear. Nod your head as it goes higher -- go ahead and cry, let it out. Look at the fear: when was the last time you felt it, a few minutes ago? Does it also awaken you at night, did it awaken you Tuesday night?

C: Yes it had to do with that problem, the same problem.

K: You were dreaming about it, weren't you?

C: Yes.

K: But you awakened with it, didn't you?

C: Yes.

K: All right now, feel it growing higher, hold on you can hold onto my hand. You're feeling it go higher and higher, it's going to get pretty tough but go ahead, let it go all the way to the top. Tell me if it gets unbearable. Go ahead and cry, and tell me if it gets unbearable. It will eventually go down all by itself in a minute or so. You don't need it, and you don't have to own their problems. I know you love your Mommy and Daddy, and that's good because you are a very fine woman, but binges aren't going to help them, is it? Do you want to get rid of it now? Do you understand now that your mind brings it on?

C: Yes...

K: Good. I want you to say to yourself, "STOP it! I don't need it, and I don't want it!" Say it aloud now...

C: STOP it! I don't need it and I don't want it!

K: Take a deep breath and hold it... exhale now, relaxing... and going deeper into relaxation now... That's it, deeper now, and take another deep breath, real nice and deep, that's it; now exhale and relax, and just go deeper... just drift off, letting it all drift far behind you. Imagine the warm sun overhead, warming your body from your neck down, and the beautiful day at the beach.

(The beach had been used with this client in imagery conditioning for relaxation.)

Let it drift far behind you and nod your head as the tension moves down. See... your mind brings it on and your mind can take it away. It's very natural for you to be tense about your Mom and Dad, but remember it doesn't do any good to binge or anything else that hurts you, does it? But it's normal to have those feelings. By the way, the reason you're constipated is those feelings of tension. It's natural to become tense when a lot of things happen at once such as when work gets rough and Mom and Dad are having problems too. A lot of people have 'down' days because of problems. You're doing remarkably well in fact, so stop beating on yourself. You are a strong fine woman, aren't you?

(Client nods.)

All right, remember that now. Let's say you begin to get tense now, imagine it... Let's go back to Tuesday night. We're going to have you see there's another way to reduce the tension. Imagine now you're awakening then, feeling tense ... nod your head as you feel the tension building

(Client nods)

Now that you know your mind brings it on, you know your mind can take it away. What can you do at this point?

C: I can say, "STOP it! I don't need this! Stop it! I don't want it!

K: And then take a nice deep breath, and do that now... hold it... and now exhale and just relax. Your husband sleeps next to you, doesn't he? And you have a good marriage, correct?

C: Yes.

K: Would he mind if you awakened him and told him to give you a nice hug?

C: No, he wouldn't mind; he wants me to.

K: Good. Imagine now, what do you say to him -- imagine you shake him and awaken him, what do you say now?

C: Give me a hug.

K: Right, "Hold me, give me a hug." Your husband is a real nice guy and you have a good relationship; he gives you a nice big hug like a bear while he's half asleep, and -- you love him, right?

(Client nods.)

You feel a nice warmth in that love, and you begin to think about the good life that you have and that you are loved. Mom and Dad love you too, but Dad's got problems. They're really not your problems, you have your own family to be concerned about, but you'll help Mom and Dad out all you can. You can't own other people's problems. The best way you can help everybody is for you to stay healthy and strong, to love yourself and to take good care of you. So the best thing you can do to help Mom especially is for you to continue getting better. It's a nice thought, isn't it? What's best for you is best for your loved ones.

C: Yes.

K: Just imagine yourself going back to sleep now and remember, if you awaken hungry, it's because your blood sugar is down, and that means to just have a little piece of something or you can just go back to sleep and have a good breakfast in the morning, OK? You know what has been awakening you now, you know what you've been reducing, and it's understandable. Now you have some choice to make when this occurs. So relax, and imagine you're going back to sleep. Rest for a few minutes now, just rest, sinking deeper and deeper, deeper into relaxation. You understand now that these are only feelings and emotions, and they will never hurt you. You know where they are coming from, you know what's happening, you know that your mind brings them on, and your mind can take them away. Now just relax...

(Client is given a five-minute rest period.)

Do you feel a little bit relieved now?

(Client nods.)

Good. Just continue relaxing now. Each time we bring these feelings up and put them down, they'll get less and less. So we're going to bring them up again. I want you to think about Mom, think about Dad, think about worrying about what's going to happen to Dad -- Dad may die an alcoholic, right? That's a horrible thought about which you're worried, right? You didn't want to worry about it, you tried not to, but it kept coming into your mind, didn't it? So face it now, it may come true, face it, feel the tension growing now, growing higher and higher. Nod your head as it grows higher and higher, higher and higher...

C: I feel like he may get killed in a car accident too, as he shouldn't be driving.
K: But your worrying about it isn't going to stop it from possibly happening, is it? It is natural to worry, but let's get rid of these conditioned fears that your Dad may die in a car accident. It would hurt Mom's feelings, right? You feel hurt, and you're angry and hurt at the same time, aren't you? Humiliated and worried because Daddy's a drunk, right? I know it hurts, Daddy's a drunk, go ahead and cry... hold on, and go ahead and cry. Daddy's a drunk, that's what hurts? You're hurt and you're angry, aren't you... You feel you should be neither, right? You should be both angry and hurt, that's natural -- go ahead, you're angry at him, aren't you? -- and hurt at the same time, broken-hearted, feeling unloved... After all, if he loved you, the guy would act a lot differently, right? That's not true, though, he's a confused man. Go ahead and keep crying, cry it out... Beginning to go down now?
C: Yes.
K: Okay, that's it... there are so many good things about your life, now. Take a good look at your life, your marriage, think about you. You're getting healthier and stronger and you're looking prettier than ever before. Everyone's telling you that, aren't they?
C: Yes.
K: Imagine, look at your Mom's face. There's one thing that is working out all right in Mom's life, and what's that? --you're getting better!
C: I am getting better, and she's glad.
K: Yes, it takes a lot of worry off of Mom. You always wanted to please her anyway, because you know she's had some rough times. You know a mother's love is usually real love, and your Mom really loves you because she was happy that you were doing well. When she would say to you, "You'd better eat, " what was she really saying to you?
C: She was trying to make me better.
K: That's right.
C: One thing that I think that might have triggered the binge was that when I was listening to the tapes, Mom called. I woke up, and I guess I was still under hypnosis, and probably the idea that she was talking about Dad...
K: --went right into your subconscious mind?
C: Yes.
K: I don't want you to get up during the hypnosis tapes to answer the phone. Unplug (or mute) it.
C: Okay.
K: Great. Okay, let's go through it again...
Think about Mom, think about the hurt and anger, Dad's a drunk, right? And that hurts, doesn't it? And calling your daughters bitches, right? That really hurt, didn't it -- what kind of father would call his own granddaughters bitches, right?
(Client nods)
Feel the hurt growing higher and higher... this time, is it anger, hurt, tension? Describe it to me.
C: Not really so much anger as fear and tension.
K: Feel the fear, feel the fear and tension. Feel the fear building higher and higher, and let's face that fear.... Feel it building higher and higher, think about it, think about Dad getting killed in a car accident,

about Mom's hurt and pain. Think about your hurt and pain... think about your daughters and Daddy calling them bitches. Nod your head as it grows higher and higher.
(Client nods.)
It's getting easier each time, isn't it... you can get to the point eventually that you won't be able to even produce it anymore if you don't want to. Are you getting tenser?
C: Yes.
K: Think about Daddy, Daddy's a drunk, Daddy's a drunk-- the humiliation, fear, sympathy, right? What other thoughts are behind all this? He's killing Mom by what he's doing, you love Mom, so you feel hurt about it since you can't control it?
(Client nods.)
You can't-- you're not God, you can't control it. Mom's pretty tough, and I think Mom's going to survive this. I think you inherited something from her, you seem to be a survivor too. Let's keep going... Daddy's a drunk and it's humiliating, right? All right, we're going to write it on the side of a blimp and fly it over the city so that all the important people can see and know that your Daddy is a boozer, okay? ……..Are you still feeling fatter now, or just tense?
C: It's mostly tension.
K: It's interesting how it all gets tied together isn't it? The tenser you got, the more you wanted to go out and binge. You don't have to hurt yourself anymore, though; you'll just tough it out instead. Think about it: Daddy's a boozer, Daddy's a drunk... What words do you use that really bother you?
C: That you were using?
T: No, those that you use that really bother you. Face up to the thoughts about Daddy.
C: Daddy's a drunk, he's irresponsible, he won't take care of himself. He won't stop drinking, he becomes so irresponsible... becomes so mean and ugly. His personality is gross.
K: A big disappointment, isn't it?
C: A very big disappointment. I feel hurt.
K: Feel the hurt, feel the big hurt... you're going to face it and get rid of it. You understand what we're doing now, we're extinguishing those feelings. So face it, don't fight it, let it overwhelm you. Think about it: he's killing Mom because of her worrying, isn't he? It breaks your heart and you feel sorry for your sister also, right?
(Client nods.)
You feel sorry for the kids, and it just breaks your heart because you're a kind woman. We're going to have to get rid of this damned hurt because it's not doing any good, it's getting in the way of your life. There's nothing you can do to help. Dad's going to have to help himself the way you've done. Feel the hurt, let it go, higher and higher, higher... it's going down, isn't it?
C: It's almost gone! (Client's tone of voice was one of surprise, although she had previously experienced abreactive extinction and in-vivo extinction.)
K: Good! It's getting easier now. Let's do it one more time now so that we won't have the bad feeling hitting you at night. One more time now: think about Daddy the drunk irresponsible, what a shame. Look at Mom, she's ruining her life, isn't she?
C: She's not, Daddy is.

K: That's right, she's not, but Daddy is. Mom seems to be a pretty tough old bird, doesn't she?

C: (Smiles.) Yes.

K: Yes, I think you inherited some of that toughness from her. Isn't that nice?

C: Well, in a way we both get it from my grandmother -- she was a tough one.

K: That's great. You feel a lot tougher now, don't you?

C: Yes.

K: Look at your father now, sympathy is fine, but hurt and pain aren't. You've got too much to lose you see, if you feel that way; if it helped him to feel that hurt and pain, I'd say do it; you'd cut off your hand to help him, wouldn't you? But you know hurting yourself isn't going to help anyone.

C: It wouldn't do any good.

K: No, he can only help himself. He's in the hands of God, and we're not God, and we have to accept that.

(The client was religious and through hypnotherapy had eliminated her fear of God and accepted the Christian doctrine concerning guilt in prior sessions.)

Remember, to own other people's problems doesn't help. I can certainly see why you've done that, though; that's human. Now, you must put first things first and that is you, because without you staying healthy and strong, you can't help anyone. Without your strength and health, you cannot satisfy responsibilities to those who are important, namely your husband and child. Your health comes first. The hurt and pain is not doing anyone any good; they're not helping you, so look at the situation and remember to have faith. Do you pray about it?

C: Yes.

K: Good. Pray for him, and that's all you can do; the rest is in the hands of God. Your Mother will survive and your children will not be tainted for the rest of their lives because of your Dad, because all the good things you and your husband have done cannot be erased by your Dad's behavior. Simple as that. So relax, he can call them bitches or anything he'd like until he's blue in the face, he can jump up and down like a monkey, and it's not going to erase all the good that you and your husband have done. All because there's love in your family, isn't there?

(Client nods.)

You've not been a perfect mother, your husband hasn't been a perfect father, because perfect parents don't exist. But you do have the most important thing, love. So don't worry about it. Relax now, relax and rest. Just imagine yourself now out on that beach on that beautiful warm day, see yourself, and take a good look at yourself. See how good you look and feel. Do you have fat thighs?

C: No.

K: No, you don't have fat thighs. That was your mind causing you to think that you did before. So take a good look at you: you look good, you look better than you have for years, don't you? Look at you, feel it, it's good to be healthy and relaxed. You're even looking a little younger, aren't you?

(Client nods.)

So, just relax and enjoy the nice view and the warm sun, as your arms and legs become loose, having nice quiet feelings within your stomach. The constipation is only due to tension. As you relax, you feel the warm feelings within your stomach, and you'll begin to have more regular bowel movements. Feel the nice cool quiet feelings over your forehead. Feel a nice smile over your face. Think about lying in bed, and awakening... if you awaken with an urge to binge, you will stop it by using your thought-stopping exercise, saying "Stop it! I don't want it, and I don't need it," and then taking a deep breath, exhaling and relaxing, imagining a nice soft summer day at the beach or pool... and then simply rolling over and awakening somebody who loves you. How does that sound? You'll get a hug from someone who really loves you, your husband really loves you. Isn't that a nice feeling?

It's a good life, you've done a fine job, and you're doing a fine job. Never lose sight of that. Listen to the birds outside, can you hear them? (Reference is to actual birds outside the office window.) That's life, you can always take that break to listen to the birds. Relax, life is a state of mind. Your life is a good one. Only God is perfect, so you cannot be a perfect parent, you cannot be a perfect daughter, and we all have to accept that we're human. There's certain things that are out of our hands and that's where faith comes in. So accept it, there's only one perfect Being, the Deity, God, so relax.

When you want to help your Mom, there's only one way, and that's to keep getting better, to keep relaxing and becoming healthy and strong. That's what is most important to her, because she loves you. It's a nice thought, isn't it? That what is most important to her is love. Real love is when someone wants the best for you; by you doing the best for you, it makes them feel good. That's a nice deal. Now what I'm going to do is awaken you. You're going to remember all that went on, and feel relieved as well as feeling good all over.
(client is awakened.)
Feel like you got rid of something?
C: Yes.
K: So now it's obvious what has been awakening you. It will be very interesting to see what happens over the next week.

At a one year follow-up, she reported only occasionally binge-eating and vomiting at night; the amount of food consumed during the binge was very small and consisted of healthy food. Her improvements in other areas such as weight and self-esteem stabilized during this period.

R. Post Traumatic Stress Disorder (PTSD)

Post traumatic stress disorder (PTSD) as a diagnosis has gained much popularity since its creation as a separate diagnostic category in 1980. PTSD is classified as an anxiety disorder and fits well within my etiological theory and treatment approach.

Trauma is any event that causes extreme anxiety, which then becomes associated with a variety of stimuli and thoughts present at the time of the trauma. The two-process theory of avoidance learning applies. The sufferers attempt to avoid the CERs and the stimuli that elicit them when they feel anxious. The whole range of extreme anxiety-mediated reactions such as depersonalization, physiological hyperarousal, dissociative reactions, intrusive negative thoughts, and flashbacks that relate to the trauma may be experienced. Nightmares and sleep disturbances are also common, as are OCD and depression. Trauma may also cause a pseudo-psychotic dissociative state labeled *post-traumatic psychosis*. The sufferer becomes non-communicative and may experience hallucinations. This state may be confused with other psychoses. 85% of those diagnosed with PTSD qualify for at least one other psychiatric diagnosis, and 50% qualify for 3 or more. 20% of the PTSD population attempt suicide, compared to 4% of people with other diagnoses.

When PTSD was formulated in 1980, an important criteria was that the traumatic event would cause significant distress in almost any person and had to be outside the range of normal experience. Of course, rape, natural disasters, severe accidents and combat were designated as significant stressors. The stressors, not the individual's emotional reactions to the stressors, were emphasized (McNally, 2008). Since the first World War, traumatized soldiers who were unable to function were labeled as victims of "combat fatigue" or "shell shock." Extreme chronic anxiety and hyperarousal do cause lasting negative effects for many, and these soldiers were not malingering. Their reactions were out of their volitional control.

In the DSM IV-TR, the range of what was considered to be traumatizing was broadened. Individuals' reactions had to involve intense fear, helplessness, and/or horror. Of course, people vary in their reactions to similar traumatic events based upon their past experiences and genetic makeup. Human reactions to traumas, the resulting anxiety and people's subsequent reactions to that anxiety are varied and complex, making diagnostic categories only a short description of their problems.

Negative intrusive thoughts and images are a major complaint of most PTSD sufferers. Trying to wilfully suppress these thoughts and associated anxiety causes them to emerge later at greater intensity (Wegner, 2002). As stated, it would be maladaptive for humans to have the ability to will away learned fear reactions. The intrusive thoughts and frightening images concerning the trauma can be viewed as a way for the nervous system to keep the victim prepared for another life-threatening experience. However, this mechanism has become overly active in PTSD.

Since ADs are based on CERs, conditioned anxiety reactions explain the same basic mechanism underlying PTSD. ADs are the result of small traumas that compound into the disorder. Whether an AD develops more obsessions, compulsions, depression or dissociative reactions depends upon their predispositions and early experiences. The same is true for PTSD sufferers.

Early traumatic interpersonal experiences create a person's orientation towards security and people in general, and often underlie their belief systems. Later traumatic experiences both physical and interpersonal can shatter pre-existing values, sense of self, and influence how people guide their lives. All of this can leave them confused and sometimes paralyzed.

Experiences that shatter one's confidence in their conception of reality and how they thought they were valued can be life shattering and cause more extreme anxiety than physical trauma. The more traumatic these experiences are and the younger the victim is, the more intense the anxiety. For example, the trauma of an eight-year-old boy (PTSD Case 2) realizing his caretakers didn't care whether he lived or died triggered separation anxiety that mediated a variety of beliefs and behaviors.

As stated, separation anxiety may be the basic anxiety experienced by humans because it means death. To an intelligent child, abandonment and hostility from his protectors means he will be preyed upon and possibly die. In PTSD Case Study 2, this is what happened. He was preyed upon physically and experienced interpersonal traumas that set the stage for his adaptable and maladaptive attitudes and behavior. It is easy to see why an adult emotionally abandoned as a child can become untrusting, defensive, and suspicious.

Some authorities try to divide traumatized clients into simple and complex. As with most classifications in mental health, particularly dichotomous ones, the divisions make little sense. Traumatizing experiences vary from mild to severe. Traumatized people are anxious and anxiety is on a continuum from mild to severe. The troublesome behaviors and thoughts they experience that get them classified are also on a continuum. All AD clients are "traumatized," meaning they are troubled by behaviors and thoughts that are mediated and reinforced by anxiety.

Simple traumas are those resulting from one exposure. If they occur after childhood, they are easier to extinguish compared to a major trauma or repeated smaller traumas experienced at a young age. Early repeated traumas have more time to mediate more complex maladaptive coping and avoidance.

For example, a person can experience a serious accident and become agoraphobic and dissociative, but not have their basic attitudes concerning their significant relationships damaged. This is described in PTSD Case I. PTSD Case 2 provides an example of repeated interpersonal traumas, helplessness, and physical traumas, which caused the person to develop a "paranoid personality." These traumas convinced him that people are unreliable, do not care about him and will turn on him in his hour of need. He concluded and accepted he was alone in a dangerous and hostile world.

In both cases, the traumatic situations similarly caused both to experience extreme helplessness. In PTSD Case 1, the young man was pinned in a car, could not escape, and in the second case the person did not feel he could get others to care about him. His repeated physical and emotional early childhood and adult traumas added to his sense of helplessness. This led him to become homicidally aggressive at times. If either of these clients would have been able to stop the helpless situations, they would have been much less traumatized.

Post Traumatic Stress Disorder (PTSD) Case Study 1

This client would be classified as a simple trauma because the traumatizing incident occurred once. Age is also important as he was 16 years old when it occurred and was almost 20 when he consulted me. He was referred by his physician for generalized anxiety, agoraphobia, and depersonalization. When he would leave his house to go to work, he could not look up at the sky, and reported that the night sky was particularly troubling. When he did look up, he became intensely anxious and experienced depersonalization. His family members reported to him that at times he appeared spacey and in a trance.

He explained these problems began a few years ago, but could not remember exactly when. When asked, he said he had had no car accidents, hospitalizations or other trauma. As with most clients in this situation, he was frightened he was losing his mind. He was working and was not deriving any secondary gain from his problems. There was no family history of psychosis or suicide. His interpersonal relationships were good, and he was not shy or particularly anxious as a child. However the anxiety was causing him to become socially isolated and housebound, other than going to work.

I explained to him that all his symptoms were a result of anxiety and guaranteed him he was not losing his mind. The highest education level obtained was high school, and he appeared to have average intelligence. He quickly became open and friendly and expressed he did not want medication. This was why he requested to see a psychologist instead of a psychiatrist.

After taking his history, I explained how hypnosis works and how it could help him. I made him a twenty-minute hypnosis audio recording which included relaxation imagery, and emphasized that anxiety, even though miserable, was a normal reaction and that it could not hurt him nor was he losing his mind. I put much emphasis on this as I do with every client. This is the standard recording I make for everyone.

He returned a week later and stated that the thought-stopping suggestion (whenever he felt anxiety he was to say to himself silently in an aggressive way, "STOP IT! I don't want this!" and to then take a deep breath, exhale, and to imagine a relaxing scene) helped to some degree. He also stated that knowing that, even though the anxiety felt horrible, it was a natural response and could not hurt him was extremely helpful. He rated his improvement at 15%. During this second meeting I explained I would regress him under hypnosis to the anxiety that was bothering him and then repeatedly bring it up in degree and then bring it down until it disappeared. This is of course abreactive extinction. I had scheduled him for two hours in case the extinction took longer than expected.

My goal was to use abreactive extinction in a controlled fashion to slowly extinguish parts of the anxiety. Again, the simple vs. complex trauma dichotomy does not apply. It does not matter whether the mediating and reinforcing anxiety was learned on one occasion or many. Anxiety is anxiety, and it is extinguishable unless dissociation gets in the way.

He readily regressed to three years earlier when he had a serious car accident where his car had gone into a gully, and he had been trapped within it. It had taken a long time before he had been found by rescue workers. He remembered looking up at the sky for hours while it turned from day to night, thinking he was dying. When the first rescue worker approached the car, he had heard him comment, "He must be dead." This phrase stuck in his mind. He experienced no permanent visible injuries, although he was initially hospitalized.

I repeatedly abreacted the anxiety while emphasizing he was alive and well. I had him elicit the anxiety, and then paired it with taking a deep breath, exhaling and relaxing; the anxiety was further reduced through these relaxation techniques. I kept emphasizing to him how this traumatic event had created his anxiety, but that his imagination can trigger as well as take the same anxiety away.

After he was awakened from the trance, he felt much better. He expressed it helped to know he had a solid reason for his anxiety, and I asked him why he never told me about his car accident. He shrugged and said he hadn't thought it was important and had not wanted to discuss it.

I explained how he had to practice in-vivo extinction and that the next time it would be much easier. By practicing this procedure on his own, it would mean he would need fewer sessions. This was motivating for him since cost was a factor to him. He practiced the in-vivo procedures over the next two weeks, and when he returned for the third session he reported he was 90% or better "improved".

Spontaneous recovery was then explained to him, and he was directed to flood himself when it occurred and make a joke out of it. He liked my explanation that the difference between a brave man and a coward is that a brave man faces his fears and does what he has to do, whereas a coward feels the fear and runs away, and/or quits. Feeling anxiety or fear is normal; how you handle it is what is important. Only a crazy person does not feel fear or learn to feel anxious in dangerous situations.

At the time I was working in the referring doctor's office, who reported to me after six months that the client was doing fine. I tell all clients that if they start having problems to return immediately. This client did not need to.

This case was completed rapidly because the client had not learned a variety of maladaptive and self-sabotaging attitudes and negative interpersonal behaviors. Other than the obvious presenting problem, he was well-adjusted, liked his job, had managerial aspirations, and no problems relating to people. In other words, the trauma had not led to the altering of his early learned adaptive attitudes and behavior.

I am in agreement with Brown and Fromm (1986) that often abreactive extinction alone does not cause long-lasting changes in anxiety. However, abreactive extinction does initially reduce the anxiety and, when coupled with relaxation practiced in-vivo, it is quite effective. Abreactive extinction gets the extinction process going. Obviously, extinguishing the anxiety in an office is going to produce some generalization to the outside world. This extinction is very helpful. Also when clients realize that using their imagination to bring their anxiety on, and then using their imagination and relaxation techniques to reduce it, causes a sense of self-efficacy as they are no longer helpless.

Post Traumatic Stress Disorder (PTSD) Case Study 2

This second case example shows how repeated early interpersonal traumas along with physical traumas help create a paranoid personality. This person had a very high IQ, was highly educated in a scientific area, and had many valid criticisms of psychotherapy and its whole philosophy. He expressed he felt that what is termed a "mental disorder" was determined by "a bunch of sissies who never shook hands with pick and shovel" and "never got hit in the face," and that "the only reason they survive is become someone else protects them. If they had to live my life, they would be working as bathroom attendants."

He had been diagnosed in his late 40s as a paranoid personality with explosive tendencies and homicidal ideation. This diagnosis was received as a result of seeing a psychiatrist one time concerning disability as a result of chronic pain and resulting sleep deficit. The psychiatrist gave him this diagnosis after he asked the client to perform a "Draw A Person" test. The client had told the psychiatrist what he thought about the validity of the test and the psychiatrist's intellectual level. This was likely construed as a threat by the psychiatrist, because the client reported the interview was ended abruptly at that point. The psychiatrist's report subsequently stated that there should not be a hearing with him because in the past, when provoked, he had become aggressive.

When I encountered him, he was receiving disability which was much less than the income he had received from his prior profession. At the time he was sparingly using narcotic medication for his pain levels. Hs work history showed he had worked since he was 14 years old, and at times had two jobs. Malingering was never alleged by his medical doctors or therapists. He normally saw another MHP because it was deemed as part of the criterion of maintaining his disability. He described that he "basically played along with my (usual) therapist," and was honest with the other MHP about this. The

MHP had also diagnosed him as a paranoid personality. When questioned about the diagnosis, the client laughed and responded, "No, all of you are f--d up, not me. However I know that thinking that helps define me as having a personality disorder."

During the interview and subsequent discussion with his other MHP, this client expressed many instances that made it clear to him that his mother did not value him and would sacrifice him to please his brothers and even cousins. For example, whenever he went to her for comfort from his earliest memory, he was quickly rejected. She greatly favored the other brother and later admitted that she was wrong and regretted it. Repeated rejection and physical beatings were encountered in the school and neighborhood environment and continued through young adulthood. Numerous instances reinforced his feelings, e.g., he walked into a bar where a robbery was occurring and someone tried to shoot him in the face.

Since trauma is anything that triggers extreme anxiety along with helplessness, separation anxiety qualifies. Separation anxiety is extremely traumatizing because it is as close as can be to an instinctual system that causes extreme fear of death. Again, when children are abandoned physically or emotionally by their caretakers, they realize and fear they may not survive.

This client was, however, able to maintain good relationships with a few people. He was married and a loving parent although he had seen a therapist some years prior to being disabled because his adolescent daughter hated him but would not tell him why. He was tolerant of this even though she never did explain her feelings; however when she reached adulthood and her attitude resurfaced, he stated he had simply stopped caring about her.

Even though therapists and friends told him that his ability to cut people out of his life and his homicidal attitudes towards those whom he perceived as a threat were not normal, he remained convinced he was right. His repeated early childhood and adulthood life experiences helped create his attitudes and concept of reality that "most will sell you for a nickel." He was convinced most MHPs were "a bunch of losers who are totally unrealistic and have destroyed society. "

This individual was not a candidate for my therapy, although he was helped somewhat with hypnotic relaxation techniques to adopt a more relaxed and positive attitude. His personality disorder was fixed as a result of repeated traumas and, as importantly, he had no desire or incentive to change.

Post Traumatic Stress Disorder (PTSD) Case Study 3

This client was a 60-year-old male who came in from out of state seeking treatment for generalized anxiety, depression, and phobias. He was married, had grown children, and ran a successful business. Some of his problems were present to varying degrees throughout his life, but about ten years prior to seeing me all of his problems had worsened.

He experienced phobias of tight shoes (dress boots in particular), hot air blowing in his face, claustrophobia, and anyone hugging him, especially from behind. Hot air in general hitting him in the face (e.g., a car dashboard heater), would cause moderate to severe panic, but the worst panic occurred if hot air were to hit him in the face from a floor register. His claustrophobia was so strong that he could not tolerate even small rooms in houses. Although he could pilot an airplane, he could not sit in the co-pilot seat; similarly, he could not ride in the passenger seat of a car or in the back seat, and always insisted on driving.

During WWII he had seen action and been a crew member on C47 transports, and he felt it all had been a great adventure. Even though he had what most would term 'traumatic' experiences during the war, he loved airplanes and flying. He had given these problems much thought over the years but had never gained insight into their origins.

Over the prior five years he had been experiencing depressive episodes where feelings of gloom and doom would just come over him. When this occurred, he would 'see' himself as a fly on a wall in a mental institution looking at himself sitting in a chair in a catatonic state. This he had insight into, however: his brother had been committed to an institution for psychosis. The brother was never normal after that. I tried to dissuade him of this fear by explaining that schizophrenia usually begins in young adulthood or earlier, and that it was rare for someone to become schizophrenic at 60 years of age. He stated that that made sense, because even as a child his brother had been strange.

He had been treated by a Freudian-type psychiatrist, but had gotten worse. He felt that the psychiatrist was implying that he was a latent homosexual or must have had a homosexual experience. I told him that Freudian theory and therapy is unscientific, and that since his psychiatrist was his age, the worst medical students in medical school at that time went into psychiatry. This seemed to help calm him, because the issue of homosexuality never came up again.

As with all clients, despite the requirement of reading my book prior to seeing me, I explained hypnosis and abreactive extinction, emphasizing that the emotions he would experience are horrible but cannot hurt him or cause him to lose his mind. The adaptive significance of being able to learn to fear specific stimuli, etc. was also emphasized. This seemed to help. My suspicion was that he did not read my book as I had instructed him to do before seeing me. (Many clients don't, so it is always necessary to review the basic concepts.)

He asked if it was important that he understand when and how he learned these anxiety responses. I presented the explanation in my book that insight into the origins of his problems is not important, and seeking or even knowing these explanations does not change the problems. What is important is that these problems are learned and can be unlearned. Furthermore, I explained that we can seek and search for years and never know how we acquired these anxieties because the human memory is basically unreliable. It is more reasonable to just get rid of the problems. This he also seemed to accept quite readily.

During the initial hypnotic induction, he seemed to be a very good subject. However, when I attempted to abreact the troublesome CERs, he immediately opened his eyes and said it was not working. Even though I explained to him again how hypnosis works, it became clear he still expected to experience some kind of altered type of state where he was unconscious. I could also detect he started to experience anxiety during the induction; when that happened, he dissociated immediately and opened his eyes.

I abandoned hypnosis at this point and explained how we were going to extinguish in-vivo each of his phobias instead. He did not want to do this, and he again needed much reassurance that encountering these anxieties would not cause him to lose his mind. Intellectually he understood this.

We secured a tight pair of dress boots and had him wear them. I then flooded him by having him focus on how uncomfortable they were and how he could not escape from them. I continually brought his attention back to his fears and negative thoughts, and flooded him for two hours. Before the 2 hours were finished, he could not emit any more anxiety and was relaxed and bored.

I then put him in the backseat of a sports car (which was very small), and drove him around while flooding him with his fears. We kept this up for approximately two hours. At the beginning of both of these sessions, he had rated his anxiety as a 10; before the two 2-hour periods were over, he rated his anxiety at zero and appeared very relaxed.

His homework assignment was to again ride around in the cramped back seat of the sports car the next morning before seeing me in the afternoon, as he had been unable to do this for over 10 years. He was able to do this successfully and extinguish his fears, and reported he had ended up enjoying the drive since he did not having the responsibility of driving.

That afternoon I described to him how we were going to next extinguish his fear of closed spaces by having him stay in a closet. After placing him in a closet with no light and closing the door, he panicked, but only minimally because he felt he could kick the door down and escape. I therefore jammed a heavy couch against the door and sat on it to keep it from moving, and asked him to try to push the door open. This triggered extreme anxiety as he could not do it. I then proceeded with the flooding while he was still in the closet by having him verbally and continually obsess about his fears aloud (such as losing his mind), and having him use the Law of Reversed Action (having him use his willpower to try to increase the fears.) He let his thoughts and fears surface, and I forced him to keep obsessing about them. After about 1.5 hours, he was over-extinguished and relaxed.

The next morning he was assigned to again ride around in the car with his relative, but this time wearing his tight boots with the hot air from the heater blowing in his face. That afternoon when we met, he reported his anxiety and fears had been extremely high at the beginning but he was able to extinguish them. I immediately drove him around with him in the front passenger seat of the sports car with the hot air blowing in his face for about another 1. 5 hours in order to see if he really had fulfilled his assignment. His anxiety started at a low level but quickly abated.

Near the end of the trip when he was most relaxed, he remarked, "You know, I wonder if this has anything to do with all of this. I was raised by a very abusive father. He beat us kids for every little thing. I remember when I was in about second grade, I was on my way to school on a rainy day, but had my brother's boots on that were too small and they were really bothering me. Somehow I got wet or peed my pants, and my groin area was wet. When I got to class, the nun hit me, put me in the clothes closet and made me stand over a heat register to dry and told me not to move. She then turned off the light and slammed the door, leaving me alone in the dark. I was paralyzed because all I could think about was my father finding out later and beating the hell out of me."

I of course agreed with him that this was at the bottom of these negative CERs and that it is interesting how his insight into the how and why came after he extinguished his fears. After returning to my office, he told me, "I know why I fear someone hugging me from behind. I was playing baseball when I was about eight years old and my father yelled for me to come home. If I wasn't home within a few minutes, he would beat me. So I dropped the baseball mitt because it belonged to another boy and started running toward home. A bigger stronger boy grabbed me, and kept saying, 'Your old man is going to beat the hell out of you!' I struggled but couldn't get free. I don't remember what happened afterwards, whether or not I was beaten." I agreed again that this must explain how he learned his fears, and that he got his wish to understand how he did learn them. There was no way to verify if his memories were true or not, of course, and it mattered little in consequence as to the success of treatment.

After one week of approximately 14 hours of therapy overall, he rated himself as 90% improved and returned home. I emphasized that he would have to continue to practice his extinction exercises in his home environment. He had a minor relapse about 4 months later. Over the phone I re-explained how extinction should have been done in his home environment and told him he could return if he liked (I strongly suspected he did not diligently practice extinction in his home environment.) Apparently he did finally practice this as he did not return; eighteen months later on follow-up he rated himself 90% better.

Cases such as this make me wonder if the behavioral approach alone would work on a significant number of cases without using regressive hypnosis and abreactive extinction. However, with resistant cases I did find that that regressive hypnosis and abreactive extinction was essential.

REFERENCES

Abramson, L., Alloy, L. & Metalski, G. (1955). Hopelessness depression. In G. Buchanan & M. Seligman, Eds. *Explanatory Style.* 113-44. NJ: Erlbaum.

Affleck, G., Tennen, H., Urrows, S. & Higgins, P. (1984). Person and contextual features of daily stress reactivity. *Journal of Personal Social Psychology:* 66 (2), 329-40.

Akiskal, H. & McKinney, W. (1975). Overview of recent research and depression. *Archives of General Psychiatry:* 32, 285-305.

Babinsky, J. & Froment, J. (1918). *Hysteria or Pithiatism.* 11, 161. London: University of London Press.

Bandura, A., Blanchard, E., & Ritter, B. (1969). Relative efficacy of desensitization and modeling approaches for inducing behavioral, affective and attitudinal changes. *Journal of Personality and Social Psychology:* 13, 173-99.

Bandura, A. (1969). *Principles of Behavior Modification.* New York: Holt, Rinehart & Winston.

Bandura, A. (1977). *Social Learning Theory.* Englewood Cliffs, New Jersey: Prentice-Hall.

Bandura, A. (1977A). Self-efficacy: toward a unifying theory of behavioral change. *Psychology Review:* 84, 2, 191-215.

Bandura, A. (1980). The self and mechanisms of agency. In J. Suls, Ed., *Social Psychological Perspectives on the Self.* Hillsdale, New Jersey: Erlbaum.

Bandura, A., Adams, N., Hardy, A. & Howells, G. (1980). Tests of the generality of self-efficacy theory. *Cognitive Therapy and Research:* 4, 39-66.

Barber, T. & Calverley, D. (1963B). "Hypnotic-like" suggestibility in children and adults. *Journal of Abnormal Social Psychology:* 66, 589-97.

Barlow, D., Ed. (2002). *Anxiety and Its Disorders, 2nd Ed.* NY: Guilford.

Barnes, B. & Galton, L. (1976) *Hypothyroidism: The Unsuspected Illness.* New York: Thomas Y. Crowell.

Barnier, A. & McConkey, K. (1998a). Posthypnotic responding: knowing when to stop helps to keep it going. *International Journal of Clinical and Experimental Hypnosis:* 46, 204-219.

Barnier, A. & McConkey, K. (1998b). Posthypnotic suggestion, amnesia, and hypnotizability. *Australian Journal of Clinical and Experimental Hypnosis:* 26, 10-18.

Beaumont, P., George, G. & Smart, D. (1976). Some personality characteristics of patients with anorexia nervosa. *British Journal of Psychiatry:* 128, 57-60.

Beck, A. (1964). Thinking and depression II: theory and therapy. *Archives of General Psychiatry:* 10, 561-71.

Beck, A. (1967). *Depression: Clinical, Experimental, and Theoretical Aspects.* New York: Hoeber.

Beck, A. (1974). The development of depression: a cognitive model. In R. J. Friedman & M. M. Katz, Eds., *The Psychology of Depression: Contemporary Theory and Research.* New York: John Wiley & Sons.

Becker, J. (1977). *Affective Disorders.* New Jersey: General Learning Press.

Bernheim, H. (1964). *Hypnosis and Suggestion in Psychotherapy: A Treatise on the Nature and Uses of Hypnotism.* (1886). Reissued with an introduction by E. R. Hilgard. NY: University Books.

Binet, A. (1896: 2012 reprint). *On Double Consciousness.* MT:Kessinger Publishing.

Bleuler, E. (1924). *Textbook of Psychiatry.* 10, 1622, 188. NY: Macmillan.

Boulougouris, J. & Bassiakos, L. (1973). Prolonged flooding in cases of obsessive-compulsive neurosis. *Behavior Research and Therapy:* 11, 227-31.

Bowlby, J. (1980). *Attachment and Loss III.* NY: Basic Books.

Breger. L. (1974). *From instinct to identity.* NJ: Prentice-Hall.

Brown, D. & Fromm, E. (1986). *Hypnotherapy and Hypnoanalysis.* NJ: Lawrence Erlbaum.

Bruch, H. (1973). *Eating Disorders.* NY: Basic Books, Inc.

Bryant. R. (2005). Hypnotic emotional numbing: a study of implicit emotion. *International Journal of Clinical and Experimental Hypnosis:* 53, 26-36.

Budzynski, T. (1978). Biofeedback and the twilight states of consciousness. In G. E. Schwartz & D. Shapiro, Eds., *Consciousness and Self-Regulation, Vol. 1.* NY: Plenum Press.

Cara, E. (7/31/2012). *The Devil You Know.* <http://web.archive.org/web/20130302063450/http://heresyclub.com/2012/07/the-devil-you-know/>.

Chaplin, J. (1974). *Dictionary of Psychology.* NY: Dell Publishing Co.

Cheraskin, E. & Ringsdorf, Jr., W. (1971). *New Hope for Incurable Diseases.* NY: Arco Publishing Co.

Clark, L. (2005). Temperament as a unifying basis for personality and psychopathology. *Journal of Abnormal Psychology:* 114, 505-521.

Crisp, A. (1980). *Anorexia Nervosa: Let Me Be.* NY: Grune & Stratton.

Dalai Lama & Cutler, H. (1998). *The Art of Happiness.* NY: Riverbend Books.

Dally, P. (1969). *Anorexia Nervosa.* London: Wm. Heinemann, Ltd.

Dare, C. & Eisler, I. (2002). Family therapy and eating disorders. In C. Fairburn & K. Brownell, Eds., *Eating Disorders and Obesity: A Comprehensive Handbook, 2nd Ed.* 314-19. NY: Guilford.

Das, J. (1958). Conditioning and hypnosis. *Journal of Experimental Psychology:* 56, 110-13.

Dawes, R. (1994). *House of Cards.* NY: The Free Press.

Delgado, J. (1969). *Physical Control of the Mind.* NY: Harper & Row.

Dessoir, M. (1890). *Erster Nachtrag zur Bibliographie des modernen Hypnotismus* (first supplement to the *Bibliography of Modern Hypnotism).* Berlin: Carl Dunckers Verlag.

Dollard, J. & Miller, N. (1950). *Personality and Psychotherapy.* NY: McGraw-Hill.

Ellenberger, H. (1972). The story of "Anna O.": a critical review with new data. *Journal of the History of the Behavioral Sciences:* 8, 267-79.

Eysenck, H. (1952). The effects of psychotherapy: an evaluation. *Journal of Consulting and Clinical Psychology:* 16, 319-24.

Eysenck, H. (1960). Classification and the problem of diagnosis. In J. J. Eysenck, Ed., *Handbook of Abnormal Psychology.* London: Pitman.

Eysenck, H. (1965). *Facts and Fiction in Psychology.* Baltimore: Penguin Books.

Eysenck, H. (1966). *The Effects of Psychotherapy.* NY: International Science Press.

Eysenck, H. (1979). Behavior therapy and the philosophers. *Beharior Research and Therapy:* 17, 511-14.

Fairburn, C. & Harrison, P. (Feb. 2003). Eating disorders. *Lancet:* 361, 407-16.

Feldman, R. & Green, K. (1967). Antecedents to behavioral fixations. *Psychology Review:* 74(4), 250-71.

Ferster, C. (1973). Behavioral approaches to depression. In R. J. Friedman & M. M. Katz, Eds., *The Psychology of Depression: Contemporary Theory and Research.* Washington, D.C.: Winston-Wiley.

Festinger, L. (1957). *Theory of Cognitive Dissonance.* Stanford: Stanford University Press.

Festinger, L. (1964). *Behavioral support for opinion change.* Public Opinion Quarterly: 28, 404-17.

Fisher, V. (1937). *An Introduction to Abnormal Psychology.* 162. NY: Macmillan.

Frankl, V. (1960). *Man's Search for Meaning: An Introduction to Logotherapy.* NY: Washington Square Press.

Frankl, V. (1960). Paradoxical intention. *American Journal of Psychotherapy:* 14, 520-35.

Franklin, M. & Foae, E. (2002). Cognitive behavioral treatments for obsessive compulsive disorders. In P. Nathan & J. Gorman, Eds., *A Guide to Treatments that Work, 2nd Ed.* 367-86. London: Oxford University Press.

Franklin, M. & Foae, E. (2007). Cognitive behavioral treatments for obsessive compulsive disorders. In P. Nathan & J. Gorman, Eds., *A Guide to Treatments that Work, 3rd Ed.* 431-446. NY: Oxford University Press.

Fredericks, Carlton. (1976). *Psychonutrition: The Diet, Vitamin, and Mineral Way to Emotional Health.* NY: Grosset & Dunlap.

Freud, S. (1957). *Five Lectures on Psychoanalysis.* Standard Edition, 9-55. London: Hogarth Press.

Freud, S. (1936). *The Problem of Anxiety.* NY: Norton.

Freud, S. & Breuer, J. (1940). On the psychical mechanisms of hysterical phenomena. In E. Jones, Ed., *Collected Papers, Vol. 1.* 24-41. London: Hogarth Press.

Garfinkel, P., Moldofsky, H. & Garner, D. (1977). Prognosis in anorexia nervosa as influenced by clinical features, treatment, and self-perception. *Journal of the Canadian Medical Association:* 117, 1041-45.

Garfinkel, P., Moldofsky, H. & Garner, D. (1980). Anorexia nervosa. *Archives of General Psychiatry:* 37.

Girden, E. & Culler, E. (1937). Conditioned responses in curarized striate muscle in dogs. *Journal of Comparative Psychology:* 23, 261-74.

Gotestam, K. & Melin, L. (1974). Covert extinction of amphetamine addiction. *Behavior Therapy:* 5, 90-92.

Gray, J. (1957). *Elements of a Two-Process Theory of Learning.* NY: Academic Press.

Gull, W. (1874). Apepsia hysterica: anorexia nervosa. *Clinical Society Transactions:* 7, 180-5.

Guyton, A. (1971). *Textbook of Medical Physiology.* Philadelphia: W. B. Saunders.

Halmi, K., Brodlund, G. & Loney, J. (1973). Prognosis in anorexia nervosa. *Annals of Internal Medicine: 78, 907-9.*

Harlow, H., Harlow, M. & Suomi, S. (1971). From thought to therapy: lessons from a primate laboratory. *American Scientist:* 59, 538-49.

Harris, J. (1998, 2009). *The Nurture Assumption.* NY: Simon & Shuster.

Hawkins, D. & Pauling, L., Eds. (1973). *Orthomolecular Psychiatry.* San Francisco: W. H. Freeman.

Hersen, M. & Detre, T. (1980). The behavioral psychotherapy of anorexia nervosa. In T. Karasu & L. Bellak, Eds., *Specialized Techniques in Individual Psychotherapy.* NY: Brunner-Mazel: 295-304.

Hilgard, E. (1965). *Hypnotic susceptibility.* NY: Harcourt, Brace & World.

Hilgard, E. (1979). Divided consciousness in hypnosis: the implications of the hidden observer. In E. Fromm & R. Shor, Eds., *Hypnosis: Developments in Research and New Perspectives.* NY: Aldine.

Himmelsbach, C. (1942). Clinical studies of drug addiction: physical dependence, withdrawal, and recovery. *Archives of Internal Medicine:* 69, 766-772.

Hirschlaff, L. (1919). *Hypnotismus und Suggestiontherapie.* Leizig: Barth.

Hobbs, N. (1962). Sources of gain in psychotherapy. *American Psychologists:* 17, 741-47.

Hodgson, R. & Rachman, S. (1972). The effects of contamination and washing in obsessional patients. *Behavior Research and Therapy:* 10, 111-17.

Hodgson, R. & Rachman, S. (1977). Obsessive-compulsive complaints. *Behavior Research and Therapy:* 15, 289-95.

Hodgson, R. & Rachman, S. & Marks, I. (1972). The treatment of chronic obsessive-compulsive neurosis: follow-up and further findings. *Behavior Research and Therapy:* 10, 181-89.

Hoffer, E. (1950). *True Believer.* NY: Harper Perennial Classics.

Homme, L. (1965). Perspectives in psychology: control of coverants, the operants of the mind. *Psychological Record:* 15, 501-11.

Hordern, A. (1952). The response of the neurotic personality to abreaction. *Journal of Mental Science:* 98, 630-9.

Horowitz, S. (1970). Strategies within hypnosis for reducing phobic behavior. *Journal of Abnormal Psychology:* 75, 104-12.

Hudson, J., Hiripie, E., Pope, H. & Kessler, R. (2007). The prevalence and correlates of eating disorders in the national co-morbidity survey replication. *Biological Psychiatry:* 61(3): 348-358.

Hurzeler, M., Gerwirtz, D. & Kleber, H. (1976). Varying clinical contexts for administering Naltrexone. In D. Julius and P. Renault, Eds., *Narcotic Antagonists: Naltrexone.* 48-66. Washington, D.C.: N.I.D.A. Research Monograph.

Inbar, J. & Lammers, J. (2012). Political diversity. In *Social and Personality Psychology Perspectives,* Sept. 2012, Vol. 7,(5) 496-503.

Ingram, R., Scott, W. & Siegle, G. (1999). Depression: social and cognitive aspects. In T. Millon & P. Blaney, Eds., *Oxford Testbook of Psychopathology.* 203-26. NY: Oxford University Press.

Jacobsen, E. (1938). *Progressive Relaxation.* Chicago: University of Chicago Press.

James, W. (1899). Automatic writing. *Proceedings of the American Society for Psychical Research:* 1, 548-64.

Janet, P. (1903). *Les Obsessions et la Psychasthénie.* Paris: Felix Alcan.

Janet, P. (1907). *Major Symptoms of Hysteria.* 161, 174. NY: Macmillan.

Kanfer, F. & Saslow, G. (1965). Behavioral analysis: an alternative to diagnostic classification. *Archives of General Psychiatry:* 12, 529-38.

Kappas, J. (1978). *Professional Hypnotism Manual.* Panorama City, CA: Panorama Publishing Co.

Keefe, F., Smith, S., Buffington, A., Gibson, J., Studts, J. & Caldwell, D. (2002). Recent advancements and future directions in the biopsychosocial assessment and treatment of arthritis. *Journal of Consulting and Clinical Psychology:* 70(3), 640-55.

Keppner, D. & Roccaforte, P. (1974). Role of the contralateral roots in spinal habituation. *Annual Journal of the Society of Neuroscience.* St. Louis, MO.

Kirsch, I., Lynn, S. & Rhue, J. (1993). Introduction to clinical hypnosis. In J. Rhue, S. Lynn, & I. Kirsch, Eds., *Handbook of Clinical Hypnosis,* 3-22. Washington, DC: American Psychological Association.

Kirsch, I. & Lynn, S. (1995). The altered state of hypnosis: changes in the theoretical landscape. *American Psychologist:* 50, 846-858.

Kirsten, G. & Robertiello, R. (1975). *Big You, Little You: Separation Therapy.* NY: Dial Press.

Kroger, W. (1977). *Clinical and Experimental Hypnosis.* Philadelphia: Lippincott.

Kroger, W. & Fezler, W. (1976). *Hypnosis and Behavior Modification: Imagery Conditioning.* Philadelphia: Lippincott.

Lacey, J., Kagan, J., Lacey, B. & Moss, H. (1963). The visceral level: situational determinants and behavioral correlates of autonomic response patterns. In P. Knapp, Ed., *Expressions of the Emotions in Man.* 161-96. NY: International Universities Press.

Lambert, K. & Kinsley, C. (2005). *Clinical Neuroscience: Neurobiological Foundations of Mental Health.* NY: Worth Publishers.

LeGrange, D. & Locke, J. (2005). The derth of psychological treatment studies for anorexia. *International Journal of Eating Disorders: 37, 79-91.*

Lasegue, C. (1873). On hysterical anorexia. *Medical Times, S. Gaz.:* 265-6.

Levy, R. & Meyer, V. (1971). Ritual prevention in obsessional patients. *Proceedings of the Royal Society of Medicine:* 64, 1115-18.

Lewinsohn, P. (1974). Clinical and theoretical aspects of depression. In K. Calhoun, H. Adams & K. Mitchell, Eds., *Innovative Methods in Psychopathology.* 63-120. NY: Wiley & Sons.

Lewinsohn, P. & Atwood, G. (1969). Depression: a clinical research approach. *Psychotherapy: Research and Practice: 6, 166-71.*

Lewinsohn, P. & Graf, M. (1973). Pleasant activities and depression. *Journal of Abnormal Psychology:* 79, 291-5.

Lewinsohn, P. & Libet, J. (1972). Pleasant events, activity schedules and depression. *Journal of Abnormal Psychology: 79, 291-5.*

Lewinsohn, P., Lobitz, C. & Wilson, S. (1973). "Sensitivity" of depressed individuals to aversive stimuli. *Journal of Abnormal Psychology: 81, 259-63*

Libet, J. & Lewinsohn, P. (1973). The concepts of social skill with special reference to the behavior of depressed persons. *Journal of Consulting and Clinical Psychology:* 40, 304-12.

Lifshitz, K. & Blair, J. (1960). The polygraphic recording of a repeated hypnotic abreaction with comments on abreaction psychotherapy. *Journal of Nervous and Mental Disease:* 130, 242-6.

Loftus, E. & Davis, D. (2006). Recovered memories. *Annual Review of Clinical Psychology:* 2, 469-498.

London, P. (1965a). Developmental experiments in hypnosis. *Journal Proj. Pers. Ass.*

London, P. (1969). *Behavior Control.* NY: Harper & Row.

London, P. & Cooper, L. (1969). Norms of hypnotic susceptibility in children. *Developmental Psychology:* 1, 113-24.

Lundgren, J., Danoff-Burg, S. & Anderson, D. (2004). Cognitive-behavior therapy for bulimia: an empirical analysis of clinical significance. *International Journal of Eating Disorders:* 35, 262-274.

MacPhillamy, P. & Lewinsohn, P. (1971). Relationship between positive reinforcement and depression. Paper presented at meeting of the *Western Psychological Association.* University of Oregon.

MacPhillamy, P. & Lewinsohn, P. (1974). Depression as a function of levels of desired and obtained pleasure. *Journal of Abnormal Psychology:* 83, 651-57.

Maier, N. (1949). *Frustration: The Study of Behavior Without a Goal.* NY: McGraw-Hill.

Maier, S. & Seligman, M. (1976). Learned helplessness: theory and evidence. *Journal of Experimental Psychology: General:* 105, 3-46.

Maltz, M. (1960). *Psychocybernetics.* NJ: Prentice-Hall.

Mandler, G. (1966). Anxiety. In D. L. Sills, Ed., *International Encyclopedia of the Social Sciences.* NY: Crowell, Collier & Hamilton.

Marks, I., Hodgson, R. & Rachman, S. (1975). Treatment of chronic obsessive-compulsive neurosis by in-vivo exposure: a two year follow-up and issues in treatment. *British Journal of Psychiatry:* 127, 349-64.

Maslach, D., Marshall, G. & Zimbardo, P. (1972). Hypnotic control of peripheral skin temperature: a case report. *Psychophysiology:* 9, 600-5.

Masserman, J. (November, 1967). The neurotic cat. *Psychology Today.*

Mather, N. (1970). The treatment of an obsessive-compulsive patient by discrimination learning and reinforcement of decision making. *Behavior Research and Therapy:* 8, 316-18.

McDougall, W. (1911). Suggestion. *Encyclopedia Brittanica:* 162.

McNally, R. (2008). Trauma in childhood (Letters to the Editor). *Archives of General Psychiatry:* 64(12), 1451.

Meares, A. (1972). *A System of Medical Hypnosis.* NY: Julian Press.

Mendels, J., Ed. (1973). *Biological Psychiatry.* NY: John Wiley & Sons.

Mendels, J. & Frazer, A. (1974). Brain biogenic amine depletion and mood. *Archives of General Psychiatry:* 30, 447-51.

Messerschmidt. R. (1933a). Response of boys between the ages of 5 and 16 years to Hull's postural suggestion test. *Journal of Genetic Psychology:* 43, 422-37.

Messerschmidt. R. (1933b). The suggestibility of boys and girls between the ages of 5 and 16 years. *Journal of Genetic Psychology:* 43, 422-37.

Meyer, F. (1966). Modification of expectations in cases with obsessional rituals. *Behavior Research and Therapy:* 273-80.

Mezey, A. & Cohen, S. (1961). The effect of depressive illness on time judgement and time experience. *Journal of Neurology, Neurosurgery, and Psychiatry:* 24, 269-70.

Mikulas, W. (1974). *Concepts in Learning.* Philadelphia: W. B. Saunders.

Miller, N. E. (1975). Applications of learning and biofeedback to psychiatry and medicine. In A. M. Freedman, H. I. Kaplan & B. J. Sadoch, Eds., *Comprehensive Textbook of Psychiatry II, Volume I, 2nd ed.* Baltimore: Williams & Wilkins.

Miller, W. E. (1959). Liberalization of basic S-R concepts: extensions to conflict behavior, motivation, and social learning. In S. Koch, Ed., *Psychology: A Study of a Science, II.* NY: McGraw-Hill.

Miller, W. R. (1975). Psychological deficit in depression. *Psychological Bulletin:* 82, 238-60.

Miller, W. R. & Seligman, M. (1975). Learned helplessness, depression, and the perception of reinforcement. *Behavior Research and Therapy:* 4, 191-7.

Mills, H., Agras, W., Barlow, D. & Mills, J. (1973). Compulsive rituals treated by response prevention. *Archives of General Psychiatry:* 28, 525-29.

Milner, G. (1963). Ascorbic acid in chronic psychiatric patients – a controlled trial. *British Journal of Psychiatry:* 109, 294-99.

Mineka, S. & Ben Hamida, S. (1998). Observational and non-conscious learning. In W. O'Donohue, Ed., *Learning and Behavior Therapy.* 421-39. MA: Allyn & Bacon.

Minuchin, S., Roman, B. & Baker, L. (1978). *Psychosomatic Families.* MA: Harvard University Press.

Morgan, J. (1936). *The Psychology of Abnormal People.* NY: Longmans.

Morgan, H. & Russell, G. (1975). Value of family background and clinical features as predictors of long term outcome in anorexia nervosa: 4 year follow-up study of 41 patients. *Psychological Medicine:* 5, 355-71.

Mowrer, O. (1947). On the dual nature of learning – a reinterpretation of "conditioning" and "problem solving". *Psychological Medicine:* 5, 355-71.

Mowrer, O. (1960). *Learning Theory and Behavior.* NY: John Wiley & Sons.

O'Brien, C. (1974). "Needle freaks" – psychological dependence on shooting up. *Medical World News Review:* 1, 35-6.

O'Brien, C., Testa, T., O'Brien, R. & Greenstein, R. (1976). Conditioning in human opiate addicts. *Pavlovian Journal of Biological Sciences:* 24, 682-5.

Orne, M. (1977). The nature of hypnosis: artifact and essence. *Journal of Abnormal and Social Psychology:* 58, 277-99.

Overmier, J. & Seligman, M. (1967). Effects of inescapable shock upon escape and avoidance learning. *Journal of Comparative and Physiological Psychology:* 63, 23-33.

Overton, D. (November, 1969). High education. *Psychology Today:* 3(6), 48-51.

Palazzoli, M. (1978). *Self-Starvation.* NY: Jason Aronson.

Pavlov, I. (1927). *Conditioned Reflexes.* London: Oxford University Press.

Perlmuter, L. & Monty, R., Eds. (1980). *Choice and Perceived Control.* Hillsdale, New Jersey: Erlbaum.

Perlmutter, D. (2013). *Grain Brain.* NY: Little, Brown.

Pike, R. & Brown, M. (1967). *Nutrition: An Integrated Analysis.* NY: Wiley.

Pike, K., Walsh, B., Vitousek, K., Wilson, G. & Brower, J. (2003). Cognitive behavioral therapy in the post-hospitalization treatment of anorexia. *American Journal of Psychiatry:* 160(11), 2046-49.

Piper, A., Lillevik, L. & Kritzer, R. (2008). What's wrong with believing in repression? *Psychology, Public Policy, and Law:* 14, 223-242.

Pope, H., Oliva, P. & Hudson, J. (2002). Repressed memories: The scientific status of research on repressed memories. In D. Fagman, Ed., *Science in the Law: Social and Behavioral Science* Issues: 487-526. St. Paul, MN: West Group.

Prince, M. (1909). Experience to determine co-conscious (subconscious) ideation. *Journal of Abnormal Psychology:* 3, 33-42.

Rachman, S. & Shafran, R. (1998). Cognitive and behavioral features of obsessive-compulsive disorder. In R. Swinson, M. Antony, S. Rachman & M. Richter, Eds., *Obsessive-Compulsive Disorder: Theory, Research, and Treatment.* 51-78. NY: Guilford.

Rahman, L., Richardson, H. & Ripley, H. (1939). Anorexia nervosa with psychiatric observations. *Psychosomatic Medicine:* 1, 335-65.

Rachman, S., Hodgson, R. & Marks, I. (1971). The treatment of chronic obsessive-compulsive neurosis. *Fehavior Research and Therapy:* 9, 237-47.

Rachman, S., Marks, I & Hodgson, R. (1973). The treatment of obsessive-compulsive neurotics by modeling and flooding in vivo. *Behaviour Research and Therapy:* 11, 465-71.

Resacorla, R. & Solomon, R. (1967). Two-process learning theory: relationships between Pavlovian conditioning and instrumental learning. *Psychological Review:* 74, 151-82.

Roden, J. (1978). Stimulus-bound behavior and biological self-regulation: feeding, obesity, and external control. In G. E. Schwartz & D. Shapiro, Eds., *Consciousness and Self-Regulation, Vol. 2.* NY: Plenum Press.

Roemer, L. & Orsillo, S. (2009). *Mindfulness- And Acceptance-Based Behavioral Therapies in Practice.* NY: Guilford.

Roper, C., Rachman S. & Hodgson, R. (1973). An experiment on obsessional checking. *Behaviour Research and Therapy:* 11, 271-77.

Rosenthal, D. (1971). *Genetics of Psychopathology.* NY: McGraw-Hill, 1971.

Rubenstein, C. (1931). Treatment of morphine addiction in tuberculosis by Pavlov's conditioning method. *American Review of Tuberculosis:* 24, 682-5.

Sabry, A., Campbell, J., Campbell, M. & Forbes, A. (1974). Nutrition Canada. *Nutrition Today:* 9, 5-13.

Sackeim, H. & Gur, R. (1978). Self-deception, self-confrontation, and consciousness. In G. E. Schwartz & D. Shapiro, Eds., *Consciousness and Self-Regulation, Vol. 2.* NY: Plenum Press.

Salter, A. (1949). *Conditioned Reflex Therapy.* New York: Farrar & Straus.

Salzman, L. (1980). *Treatment of Obsessive Personality.* NY: Jason Aronson.

Sapolsky, R. (2000). Glucocorticoids and hippocampal atrophy in neuropsychiatric disorders. *Archives of General Psychiatry:* 57, 925-35.

Satow, L. (1923). *Hypnotism and Suggestion.* London: Dodd, Mead.

Schaffer, L. (1936). *The Psychology of Abnormal People.* NY: Longmans.

Schaffer, M. & Lewinsohn, P. (1971). Interpersonal behavior in the home of depressed versus non-depressed psychiatric and normal controls: a test of several hypotheses. Paper presented at meeting of the *Western Psychological Assoc.* Mimeo: University of Oregon.

Schildkraut, J. (1965). The catecholamine hypothesis of affective disorders: a review of supporting evidence. *American Journal of Psychiatry:* 122, 509-22.

Schwartz G. & Shapiro, D., Eds (1978). *Consciousness and Self-Regulation, Vols. 1 & 2.* NY: Plenum Press.

Schwartz, J., Stoessel, P., Baxter, L., Martin, K. & Phelps, M. (1996). Systematic changes in cerebral glucose metabolic rate after successful behavior modification treatment of obsessive-compulsive disorder. *Archives of General Psychiatry:* 53, 109-113.

Schwartz, R. (1994). *Internal Family Systems Therapy.* NY: Guilford.

Secunda, S. (1973). Special Report: The Depressive Disorders. *National Institute of Mental Health.*

Segal, Z., Williams, J. & Teasdale, J. (2002). *Mindfulness Based Cognitive Therapy for Depression.* NY: Guilford.

Seligman, M. (1971). Phobias and preparedness. *Behavior Therapy:* 2, 307-20.

Seligman, M. (June, 1973). Fall into helplessness. *Psychology Today.*

Seligman, M. (1974). Depression and learned helplessness. In R. Friedman & M. Katz, Eds., *The Psychology of Depression.* Washington, DC: Hemisphere.

Seligman, M. (1975). *Helplessness: On Depression, Development and Death.* San Francisco: Freeman.

Seligman, M., Maier, S. & Geer, J. (1968). Alleviation of learned helplessness in the dog. *Journal of Abnormal Psychology,* 73(3), 256-62.

Selye, H. (1956). *The Stress of Life.* NY: McGraw Hill.

Selye, H. (1976A). *Stress in Health and Disease.* MA: Butterworth.

Shapiro, A. (1971). Placebo effects in medicine, psychotherapy, and psychoanalysis. In A. Bergin & S. Garfield, Eds., *Handbook of Psychotherapy and Behavior Change: Empirical Analysis.* NY: Wiley.

Shulman, R. (1967). A survey of vitamin B-12 deficiency in an elderly psychiatric population. *British Journal of Psychiatry:* 113, 241-51.

Simpson, H. & Leibowitz, M. (2006). Best practice in treating obsessive-compulsive disorder. In B. Rothbaum, Ed., *Pathological Anxiety: Emotional Processing.* 147-165. NY: Guilford.

Skinner, B. (1938). *The Behavior of Organisms.* NY: Appleton Century Crofts.

Slade, P. & Russell, G. (1973). Awareness of body dimensions in anorexia nervosa: cross-sectional and longitudinal studies. *Psychological Medicine:* 3, 188-89.

Slater, E. & Glithero, E. (1965). A follow-up of patients diagnosed as suffering from hysteria. *Journal of Psychosomatic Research:* 9, 9-15.

Solomon, R. (1964). Punishment. *American Psychologist:* 19, 239-53.

Solomon, R., Kamin, L. & Wynne, L. (1953). Traumatic avoidance learning: the outcomes of several extinction procedures with dogs. *Journal of Abnormal and Social Psychology:* 48, 291-302.

Spiegel, H. & Spiegel, D. (1978). *Trance and Treatment: Clinical Uses of Hypnosis.* NY: Basic Books.

Stampfl, T. & Levis, D. (1967). The essentials of implosive therapy: a learning theory based on psychodynamic behavioral therapy. *Journal of Abnormal Psychology:* 496-503.

Stampfl, T. & Levis, D. (1969). Learning theory: an aid to dynamic therapeutic practice. In L. D. Eron & R. Callahan, Eds., *The Relation of Theory to Practice in Psychotherapy.* Chicago: Aldine.

Stampfl, T. & Levis, D. (1973). *Implosive Therapy: Theory and Technique.* NJ: General Learning Press.

Steketee, G. (1993). *Treatment of Obsessive-Compulsive Disorder.* NY: Guilford.

Stoyva, J. Barber, Tl, Dicaria, L., Kamiya, J., Miller, N. & Shapiro, D. (1971-78). *Biofeedback and Self-Control.* NY: Aldine.

Stukat, K. (1958). *Suggestibility: A Factorial and Experimental Analysis.* Stockholm: Almquist & Wisell.

Sullivan, H. (1947). *Conception of Modern Psychiatry.* NY: W. W. Norton.

Sullivan, H. (1953). *The Interpersonal Theory of Psychiatry.* NY: Norton

Thoma, H. (1967). *Anorexia Nervosa.* NY: International Universities Press.

Thoreau. H. (1906/1981). *The Writings of Henry David Thoreau, 1906.* Republished as *Works of Henry David Thoreau. NY:* Avenel Books.

Walton, D. & Mather, M. (1963). The application of learning principles to the treatment of obsessive-compulsive states in the acute and chronic phases of illness. *Behaviour Research and Therapy:* 1, 163-74.

Watkins, J. (1971). The affect bridge: a hypnoanalytic technique. *International Journal of Clinical and Experimental Hypnosis:* 19, 1, 21-7.

Watkins, J. & Watkins, H. (1979). The theory and practice of ego state therapy. In H. Grayson, Ed., *Short Term Approaches to Psychotherapy.* NY: National Institute for the Psychotherapies and Human Sciences Press.

Watson, D. (2005). Rethinking the mood and anxiety disorders. *Journal of Abnormal Psychology:* 114(4): 522-536.

Wegner, D. (1994). Ironic process of mental control. *Psychology Review:* 101, 34-52.

Wegner, D. (2002). *The Illusion of Conscious Will.* MA: MIT Press.

Wegner, D. & Erskine, J. (2003). Voluntary involuntariness: Thought suppression and the regulation of the experience of will. *Consciousness and Cognition:* 12, 684-694.

Wegner, D., Schneider, D., Carter, S. & White, T. (1987). Paradoxical effects of thought suppression. *Journal of Personality & Social Psychology:* 53, 513.

Weiss, J., Glazier, H. & Pohoreck, L. (1976). Coping behavior and neurochemical changes in rats: an alternative explanation for the original "learned helplessness" experiments. In G. Serban & A. Kling, Eds., *Animal Models in Human Psychobiology.* NY: Plenum Press.

Whitlock, F. (1957). The aetiology of hysteria. *Acta Psychiatrica Scandinavica:* 43, 144-62.

Wikler, A. (1973). Dynamics of drug dependence: implications of a conditioning theory for research and treatment. In S. Fischer & A. J. Freedman, Eds., *Opiate Addiction: Origins and Treatment.* NY: Wiley.

Wikler, A. & Pescor, F. (1970). Persistence of "relapse tendencies" of rats previously made physically dependent on morphine. *Psychopharmacologia:* 16, 14-31.

Wilkens, W. (1971). Desensitization: social and cognitive factors underlying the effectiveness of Wolpe's procedure. *Psychological Bulletin:* 76, 311-17.

Williams, R. (1976). *You Are Extraordinary.* NY: Pyramid Publications.

Wilson, G. (1982). The relationship of learning theories to behavioral therapies: problems, prospects, and preferences. In J. Boulougouris, Ed., *Learning Theory Approaches to Psychiatry.* NY: Wiley.

Wilson, G. & Fairburn, C. (1993). Cognitive treatments for eating disorders. *Journal of Counseling Clinical Psychology:* 61(2), 261-69.

Wilson, G. & Fairburn, C. (1998). Treatments for eating disorders. In P. Nathan & J. Gorman, Eds., *A Guide to Treatments That Work.* 501-530. NY: Oxford University Press.

Wilson, G. & Fairburn, C. (2007). Treatments for eating disorders. In P. Nathan & J. Gorman, Eds., *A Guide to Treatments That Work* 579-610. NY: Oxford University Press.

Wolpe, J. (1958). *Psychotherapy by Reciprocal Inhibition.* Stanford University Press.

Made in the USA
Columbia, SC
24 October 2017